Ecological Networks and Greenways
Concept, Design, Implementation

The establishment of ecological networks in Europe and greenways in America has required some of the most advanced applications of the principles of landscape ecology to land use planning. This book provides a thorough overview of the recent developments in this emerging field, combining the theoretical concepts of landscape ecology with the actual practice of landscape planning and management. In addition to biological and physical considerations important to biodiversity protection and restoration, equal weight is given to cultural and aesthetic issues to illustrate how sympathetic, sustainable land use policies can be implemented. Species models and landscape models provide the link between theory and practice, with emphasis on the choice of appropriate parameters and design for the adequate translation and implementation of the models to real life situations. Examples are given for large-scale areas (Estonia and Florida) as well as regional areas such as Milan, Chicago and the Argentinian Yungas. This invaluable book will provide a wealth of information for anyone concerned with biodiversity conservation through networks and greenways and their relevance to the planning process, whether researcher, land manager or policy maker.

ROB JONGMAN is a Senior Research Fellow at Wageningen Research Institute Alterra, the Netherlands. He is a landscape ecologist with world-wide experience in co-ordinating large-scale landscape analysis and planning projects, mainly for ecological networks and river catchments. He is treasurer of the International Association for Landscape Ecology (IALE) and co-editor of *Data Analysis in Community and Landscape Ecology* (1995).

GLORIA PUNGETTI is a lecturer in landscape at the University of Reading, and a member of Darwin College, University of Cambridge. She has wide experience as a landscape researcher and consultant for governmental organisations, and as the co-ordinator of projects with transdisciplinary and international characteristics. She is active in IALE-Europe and IUCN, and co-author of *Ecological Landscape Design and Planning: the Mediterranean Context* (1999).

Cambridge Studies in Landscape Ecology

Series Editors

Professor John Wiens
Colorado State University

Dr Richard Hobbs
Murdoch University, Western Australia

Dr Peter Dennis
Macaulay Land Use Research Institute

Dr Bruce Milne
University of New Mexico

Dr Lenore Fahrig
Carleton University

Dr Joan Nassauer
University of Michigan

Dr Marie-Josee Fortin
University of Toronto

Professor Paul Opdam
Alterra, Wageningen

Cambridge Studies in Landscape Ecology presents sympathetic and comprehensive examinations of topics that reflect the breadth of the discipline of landscape ecology. Landscape ecology deals with the development and changes in the spatial structure of landscapes and their ecological consequences. Because humans are so tightly tied to landscapes, the science explicitly includes human actions as both causes and consequences of landscape patterns. The focus is on spatial relationships at a variety of scales, in both natural and highly modified landscapes, on the factors that create landscape patterns, and on the influences of landscape structure on the functioning of ecological systems and their management. Some books in the series develop theoretical or methodological approaches to studying landscapes, while others deal more directly with the effects of landscape spatial patterns on population dynamics, community structure, or ecosystem processes. Still others examine the interplay between landscapes and human societies and cultures.

The series is aimed at advanced undergraduates, graduate students, researchers and teachers, resource and land-use managers, and practitioners in other sciences that deal with landscapes.

The series is published in collaboration with the International Association for Landscape Ecology (IALE), which has Chapters in over 50 countries. IALE aims to develop landscape ecology as the scientific basis for the analysis, planning, and management of landscapes throughout the world. The organisation advances international cooperation and interdisciplinary synthesis through scientific, scholarly, educational, and communication activities.

Also in Series

Integrating Landscape Ecology into Natural Resource Management (2002)
EDITED BY JIANGUO LIU AND WILLIAM W. TAYLOR
0521 78015 2 (hardback)
0521 78433 6 (paperback).

EDITED BY

ROB H. G. JONGMAN
WAGENINGEN RESEARCH INSTITUTE ALTERRA

GLORIA PUNGETTI
UNIVERSITY OF READING
UNIVERSITY OF CAMBRIDGE

Ecological Networks and Greenways
Concept, Design, Implementation

170301

CAMBRIDGE
UNIVERSITY PRESS

PUBLISHED BY THE PRESS SYNDICATE OF THE UNIVERSITY OF CAMBRIDGE
The Pitt Building, Trumpington Street, Cambridge, United Kingdom

CAMBRIDGE UNIVERSITY PRESS
The Edinburgh Building, Cambridge CB2 2RU, UK
40 West 20th Street, New York, NY 10011–4211, USA
477 Williamstown Road, Port Melbourne, VIC 3207, Australia
Ruiz de Alarcón 13, 28014 Madrid, Spain
Dock House, The Waterfront, Cape Town 8001, South Africa

http://www.cambridge.org

First published 2004

Printed in the United Kingdom at the University Press, Cambridge

Typeface Lexicon 10/14 pt.   *System* LaTeX 2_ε [TB]

A catalogue record for this book is available from the British Library

Library of Congress Cataloguing in Publication data

Ecological networks and greenways: concept, design implementation / edited by
Rob H. G. Jongman and Gloria Pungetti.
    p.   cm. – (Cambridge studies in landscape ecology)
    Includes bibliographical references (p.).
    ISBN 0-521-82776-0 (hc.) – ISBN 0 521 53502 6 (pbk.)
    1. Greenways.   2. Urban ecology.   3. Landscape ecology.   4. Landscape protection.
I. Jongman, R. H.   II. Pungetti, Gloria.   III. Series.
HT241.E37 2003
304.2–dc21   2002045519

ISBN 0 521 82776 0 hardback
ISBN 0 521 53502 6 paperback

Contents

Contributors

Jack L. F. Ahern (landscape planning)
Department of Landscape Architecture
  and Regional Planning
University of Massachusetts,
Amherst, MA 01003, USA
email: jfa@larp.umass.edu

José M. Arenas (public use and
  communication)
Technical Office of the Guadiamar Green
  Corridor
Regional Ministry for Environment. Region
  Andalusia
Avda. de la Innovación s/n. Edificio Minister
41.020-Sevilla, Spain
email: otcv.natura@cma.junta-andalucia.es

María Bettina Aued (spatial ecology
  and conservation)
National Parks of Argentina
Av. Santa Fe 690
1059 Capital Federal, Argentina
email: mbaued@hotmail.com

Lía Bachman (nature conservation and land
  use)
Institute of Geography
University of Buenos Aires
Puán 480
1406 Capital Federal, Argentina
email: lbachman@filo.uba.ar

Marco Baietto (conservation biology)
Department of Environmental
  Sciences
University of Milano Bicocca
Piazza della Scienza 1
20126 Milano, Italy
email: marco.baietto@unimib.it

Luciano Bani (conservation biology, GIS
  modelling)
Department of Environmental Sciences
University of Milano Bicocca
Piazza della Scienza 1
20126 Milano, Italy
email: luciano.bani@unimib.it

Marius Bolck (nature development, rural
  planning)
Province of Gelderland
Postbox 9090
6800 GX Arnhem, the Netherlands
email: m.bolck@prv.gelderland.nl

Luciana Bottoni (conservation biology,
  zoology)
Department of Environmental Sciences
University of Milano Bicocca
Piazza della Scienza 1
20126 Milano, Italy
email: luciana.bottoni@unimib.it

Irene M. Bouwma (forestry, nature
    conservation planning)
European Centre for Nature Conservation
Post Box 1352
5400 BJ Tilburg, the Netherlands
Present address:
Wageningen UR
Alterra Green World Research
Post Box 47
6700 AA Wageningen, the Netherlands
email: irene.bouwma@wur.nl

Margaret (Peggy) H. Carr (nature conservation
    and recreation planning)
Department of Landscape Architecture
P.O. Box 115704
University of Florida
Gainesville, FL 32611, USA
email: mcarr@ufl.edu

Giuseppe De Togni (urban planning)
Province of Bologna
Strada Maggiore
80 - 40125 Bologna, Italy
email: pampa@provincia.bologna.it

Donna L. Erickson (landscape planning)
School of Natural Resources and Environment
University of Michigan
430 E. University Ave.
Ann Arbor, MI 48109-1115, USA
email: dle@umich.edu

Ruud P. B. Foppen (ornithology)
SOVON vogelonderzoek Nederland
Rijksstraatweg 178
6573 DG Beek-Ubbergen,
    the Netherlands
email: ruud.foppen@sovon.nl

Thomas S. Hoctor (landscape ecology)
Department of Wildlife Ecology
    and Conservation
University of Florida
Gainesville, FL 32611, USA
email: tomh@geoplan.ufl.edu

Rinus (C.) F. Jaarsma (traffic planning)
Land Use Planning Group
Department of Environmental Sciences
Wageningen University
Gen. Foulkesweg 13
6703 BJ Wageningen, the Netherlands
email: rinus.jaarsma@wur.nl

Rob H. G. Jongman (landscape ecology)
Wageningen UR
Alterra Green World Research
Post Box 47
6700 AA Wageningen, the Netherlands
email: rob.jongman@wur.nl

Mart Külvik (landscape ecology)
Environmental Protection Institute
Estonian Agricultural University
4 Akadeemia St.
51003 Tartu, Estonia
email: mart@envinst.ee

David S. Maehr (conservation biology)
University of Kentucky
Department of Forestry
205 Cooper Bldg.
Lexington, KY 40546-0073, USA
email: dmaehr@uky.edu

Ülo Mander (landscape ecology, landscape
    planning, ecological engineering)
Institute of Geography
University of Tartu
46 Vanemuise St.
51014 Tartu, Estonia
email: mander@ut.ee

Renato Massa (ornithology, conservation
    biology)
Department of Environmental Sciences
University of Milano Bicocca
Piazza della Scienza 1
20126 Milano, Italy
email: renato.massa@unimib.it

Antonio Mora (environmental quality)
Technical Office of the Guadiamar Green
    Corridor
Regional Ministry for Environment, Region
    Andalusia
Avda. de la Innovación s/n. Edificio Minister
41.020-Sevilla, Spain
email: antonio.mora.ext@juntadeandalucia.es

Alexander J. F. M. van Opstal (ecology)
National Reference Centre for Agriculture,
    Nature and Fisheries
Ministry of ANF
Post Box 30
6700 AA Wageningen, the Netherlands
email: a.j.f.m.van.opstal@eclnv.agro.nl

Emilio Padoa Schioppa (conservation
    biology, landscape ecology)
Department of Environmental Sciences
University of Milano Bicocca
Piazza della Scienza 1
20126 Milano, Italy
email: emilio.padoaschioppa@unimib.it

Rogier Pouwels (ecological modelling)
Alterra Green World Research
Wageningen UR
PO Box 47
6700 AA Wageningen, the Netherlands
email: rogier.pouwels@wur.nl

Gloria Pungetti (landscape planning)
Darwin College
Silver Street
Cambridge CB3 9EU
email: gp114@cam.ac.uk

Kalle Remm (GIS modelling, ecology)
Institute of Geography
University of Tartu
46 Vanemuise St.
51014 Tartu, Estonia
email: kalle_r@ut.ee

Bernardino Romano (urban planning)
DAU
Department of Architecture and Planning
University of L'Aquila
Monteluco di Roio
67100 L'Aquila, Italy
email: romano@dau.ing.univaq.it

Denis A. Saunders
C/-CSIRO Sustainable Ecosystems
GPO Box 284
Canberra ACT 2601, Australia
email: denis.saunders@csiro.au

Kalev Sepp (landscape ecology)
Environmental Protection Institute
Estonian Agricultural University
4 Akadeemia St.
51003 Tartu, Estonia
email: sepp@envinst.ee

Theo van der Sluis (landscape planning, GIS
    modelling)
Wageningen UR
Alterra Green World Research
Post Box 47
6700 AA Wageningen, the Netherlands
email: theo.vandersluis@wur.nl

Daniel J. Smith (conservation biology)
Program in Landscape Ecology
Department of Wildlife Ecology
    and Conservation
University of Florida
Gainesville, FL 32611, USA
email: djs3@ufl.edu

Daniel Somma (nature conservation
    planning)
National Parks of Argentina
Av. Santa Fe 690
1059 Capital Federal, Argentina
email: djsomma@yahoo.com.ar,
djsomma@starmedia.com

Jana Verboom (ecological modelling)
Wageningen UR
Alterra Green World Research
PO Box 47
6700 AA Wageningen, the Netherlands
email: jana.verboom@wur.nl

Paul D. Zwick (GIS modelling)
Department of Urban and Regional Planning
College of Design, Construction and Planning
University of Florida
Gainesville, FL 32611, USA
email: paul@geoplan.ufl.edu

Foreword

There is an increasing trend towards urbanisation all over the world. In developed countries, around 80% of the human population now lives in cities or towns. Similarly, in developing countries, there is increasing movement to cities and towns with associated development pressures and ecological impacts in urban areas. In addition, while many rural areas are depopulating, changing land uses are placing major ecological pressures on these landscapes. These changes have had major impacts on biodiversity with detrimental changes to ecosystem processes and functions, habitat loss and fragmentation, and widespread loss of species.

Humans are a dominant species; landscape planning and management are necessary to minimise their impacts. In the past, we have relied on dedicated conservation areas and national parks to conserve the biota. We now know that these areas are usually left over after the more productive soils and landscapes have been developed for human enterprises. As a result, these areas are inadequate for the conservation purposes for which they were dedicated. The land outside these areas is also critically important for the conservation of biodiversity, and a layer of management that addresses this function now needs to be added to the more traditional management of much of the world.

Landscape connectivity is one major landscape attribute essential for the conservation of biodiversity. If we are to retain all biotic elements in landscapes and preserve ecological functions, we need to preserve the ecological connectivity of those landscapes. This involved recognising the ecological connectivity of urban and rural landscapes and planning to retain such connectivity or enhance it where appropriate. This is a major challenge for landscape ecology, that of providing practical applications to address the conservation of biodiversity across a wide range of scales from continents and regions to local districts. However, these applications must be underpinned by theoretical understanding of the functioning of ecosystems and populations and the impacts that changing land uses will have on these.

Greenways, ecological networks, landscape linkages, and corridors represent an important and critical step in the application of landscape ecology to landscape planning for conservation of biodiversity. Such ecological linkages provide advantages in terms of movement of species and nutrients, in addition to ameliorating wind and water fluxes across landscapes. They also provide a focus to galvanise social and political support to integrate human communities into landscape planning and management.

There is considerable uncertainty inherent in landscape planning and management; however, the application of ecological linkages offers manifold opportunities to develop partnerships between individuals and organisations to manage landscapes for a range of objectives, including nature conservation. We know that planning and management are experiments to which we do not know all the long-term solutions. However, the application of ecological networks and greenways offers us an adaptive management approach to our planning and actions. This approach will result not only in the development of ecological networks, but also social and political networks that will assist in the connections of people with nature conservation.

This book on theoretical and practical aspects of the design and implementation of ecological networks and greenways is an important addition to the information needed for managing natural resources and minimising our ecological footprint.

Denis A. Saunders
CSIRO Sustainable Ecosystems
Canberra, Australia

Preface

> MacArthur and Wilson's theory was and to a large extent remains
> a radical departure from mainstream thinking in contemporary
> community ecology. In its fundamental assumption it is a neutral
> theory that asserts that island communities are dispersal assembled
> and not niche assembled.

This statement was made in 2001 by Stephen Hubbell in his 'unified neutral theory of biodiversity and biogeography' (Hubbell 2002). He continues on metapopulation perspectives, asserting that this theory is very well applicable in cases of habitat fragmentation. Despite some critical discussion in scientific journals, the theory found application in planning and did change attitudes to nature conservation. The wider countryside has slowly been included in nature conservation, as also has landscape planning and land use planning.

In 1987 the President's Commission on Americans Outdoors in the United States of America recommended the '*greenways*' as new tools 'to provide people with access to open spaces close to where they live, and to link together the rural and urban spaces in the American landscape' (President's Commission 1987).

At the international conference 'Conserving Europe's Natural Heritage: Towards a European Ecological Network' held in Maastricht in 1993, Graham Bennett envisaged the need for an operational framework for guiding the implementation of strategies on European nature conservation, indicating the concept of '*ecological network*' as a tool for this (Bennett, 1994a). Engendered by the need to conserve and enhance the functioning of the ecological infrastructure of a region, the concept quickly moved on to conserving biological and landscape diversity, and to assisting other policy sectors with responsibility for sustainability and the conservation of natural ecosystems and biodiversity.

The development of ecological network and greenway concepts has been fast and they are widely used in scientific and planning literature. This book addresses both concepts and their development in two continents: Europe (EU and eastern Europe) and the Americas (North and South).

The idea for the book was generated at the world congress of the International Association for Landscape Ecology (IALE) in 1999 in Colorado where most of the authors were meeting for a symposium on ecological networks and greenways organised by the editors. In planning *Ecological Networks and Greenways* we wanted to tackle questions of concepts, design and implementation, specifically on how *econets* (ecological networks) and *greenways* can be developed on the ground after the theoretical basis has been established.

In bringing together the contributions into this book, we focused on different groups of users and we faced two main aspects of developing econets and greenways. The first aspect is the recognition that understanding the common theoretical basis of econets and greenways provides scientific knowledge that can be conveyed not only to other scientists, but also to the practitioners and politicians responsible for future implementation. In fact furthering the discourse on concepts, philosophy, methodology and history clarifies the theory and facilitates converting knowledge into practical application.

The second aspect is the requirement that we should, by illustrating the process of constructing econets and greenways, demonstrate the means of their development in the field. Thus examples of design, planning and implementation of the two subjects in different continents are essential for the necessary flow of information on more practical solutions.

The two parts of the book follow these two directions, with specification and examples in each chapter. References have been amalgamated at the end of the book, to avoid repetition and guide the reader with a single comprehensive list.

Rob Jongman and Gloria Pungetti have co-operated on the subject of ecological networks in Europe since 1996, while working together at the ECNC (European Centre for Nature Conservation) in the Netherlands. Gloria's research on ecological networks started that year after a request from ANPA (National Agency for the Protection of the Environment), Rome, to initiate the ANPA Ecological Network Action Plan with a European overview on ecological corridors, and at the same time to promote the dialogue on ecological networks in Italy. Gloria Pungetti, European Co-ordinator of the Life ECOnet Project, is a scholar of holistic landscape research and ecological landscape planning in Europe, Italy and the Mediterranean, with a focus on the relationship between nature and culture. Her professional experience at Cambridge and Reading Universities as environmental researcher and consultant for governmental organisations and local authorities is based on uniting academic theory with technical, political and social practice. Rob Jongman is a landscape ecologist trained as a vegetation scientist and working at Wageningen University and Research Centre. He specialised first in ecological monitoring and data analysis. In 1989 he started on the development of the concept and implementation of ecological networks at the European level as well as at the national and regional level. In the period 1994–97 he was based at ECNC and co-ordinated European activities on ecological networks in the framework of the Pan-European Biological and Landscape Diversity Strategy as a European approach to biological diversity.

This book contains contributions from Europe and America. We do not pretend to have included all new scientific and planning approaches. We were not able to invite all pioneers, such as the Czech and Slovak researchers and planners, nor the first planners of greenways in the USA, Canada and Australia. However, the discussion presented here and the examples of planning and implementing ecological networks show the potentials of econet and greenway development, present a broad overview and offer a challenge for authorities to renew their policy for biodiversity conservation and rural planning.

We live in an informatics world where everything changes rapidly. When the econet debate started in Italy in 1996, for example, we encountered much scepticism, but at the conference organised in 2001 for the Life ECOnet Project we found great support and recognition. Ecological networks in Europe and greenways in America are today among the most advanced applications of landscape ecological principles in land use planning. These concepts are accepted in biological and landscape conservation, not only by scientists but also by planners and political thinkers.

Furthermore the concept requires new avenues to explore; hence we are proposing in this book an approach based on both ecological and cultural linkages. We also think that in the planning of ecological networks, integration of social and ecological aspects is necessary. Nature conservation and land use planning are social actions. We show this in concrete examples from two continents and several countries: the USA, Argentina, Spain, Italy, the Netherlands and Estonia.

This book was assembled to introduce a better comprehension of econets and greenways to both scientific and planning audiences. Starting from scholars and academics, we want to exchange our experiences also with practitioners, technicians, public officers, decision makers and, last but not least, local populations. Our aim is to improve understanding, stimulate debate and encourage sustainable policies and mutual co-operation. To this purpose we found it essential to bring together theory, design, planning and practice. Although we believe in the importance of words, we also firmly believe that without sound implementation words alone will have no effect. We therefore hand over the concepts of ecological networks and greenways to all those willing to implement them. Or, as Bellamy (1994) stated at the EECONET conference in Maastricht in 1993:

> Ladies and gentlemen, time is running out for much of the biological diversity of Europe. The biological diversity of Europe is in your hands! Action must be taken now.

Acknowledgements

Ecological networks and greenways have recently been a relevant research topic among the scientists of the IALE (International Association of Landscape Ecology). The IALE World Congress of 1999 was a unique opportunity to exchange experience on the topic and develop the idea of this book. Our gratitude goes not only to IALE but also to the staff of Cambridge University Press, especially to Alan Crowden, who is active in both organisations.

This book would have been impossible without the knowledge and expertise of our 33 contributing authors, coming not only from academia but also from research institutes, governmental bodies and local organisations. We would like to express our sincere appreciation also to the reviewers of the chapters for their valuable comments and to Donald Nicolson for his expert assistance in reading the text with us.

The concepts presented here have been developed during years of study by all the authors. The editors thank especially the institutions that supported their research during this period, namely the European Commission, ECNC (European Centre for Nature Conservation), CNR (Italian National Research Council), ANPA & ARPA (Italian National & Regional Environmental Agencies), the Leverhulme Trust, the Universities of Wageningen, Cambridge and Reading, Edinburgh College of Art, and Alterra (Green World Research, Wageningen).

We also would like to thank, together with Marius Bolk, Giuseppe de Togni, Bernardino Romano and Theo van der Sluis, the EU Life Environment Programme that supported the *Life ECOnet Project* (LIFE99 ENV/UK/000177), a joint initiative of local authorities, environmental organisations and research centres from UK, Italy and the Netherlands to promote sustainability through the use of ecological networks.

Gloria was fortunate in her continued membership of Darwin College, University of Cambridge, benefiting from its stimulating multicultural and multidisciplinary environment shown as its best in the Darwin Lecture Series (published by Cambridge University Press). She is indebted to the 'college family' and to the members of the Darwin College Society for their friendship.

Daniel Smith would like to thank Joe Schaefer and Larry Harris for their support of his research and also for useful discussion and comments regarding the manuscript.

Irene Bouwma, Ruud Foppen and Alexander van Opstal would like to thank Jan van Kalkhoven and Joe P. Dirksen who participated in the research that formed the basis of their chapter. The research was made possible by a financial contribution in the framework of the Research Program of the Ministry of Agriculture, Nature Management and Fisheries, the Council of Europe and ECNC.

Kalle Remm, Mart Külvik, Ülo Mander and Kalev Sepp are grateful to the ECNC, which supported their study within the framework of the project *Indicative map of PEEN for Central and Eastern Europe*.

Rinus Jaarsma is grateful to Donna Isaac and Geert Willems, both research fellows at the Land Use Planning Group of Wageningen University, for their critical comments on his chapter, and to the anonymous referees for their contribution to improving it.

Donna Erickson appreciates funding from the *National Park Service Rivers, Trails, and Conservation Assistance* (RTCA) Program and the School of Natural Resources and Environment at the University of Michigan. In addition, her work has been improved by collaborations with Barbara Nelson-Jameson at RTCA and Norm Cox at the Greenway Collaborative, and by research assistance from Anneke Hagen, Rick Meader and Karla Rogers.

Daniel Somma, María Bettina Aued and Lía Bachman would like to thank the Argentine National Parks Administration for its regional infrastructure, the Calilegua NP staff for their knowledge and logistic backing, and the Baritu NP personnel that helped in the Oran Department searches. They also wish to thank Pablo Perovic and Sandra Caziani for their substantive field experience and comments; Jorge Wallberg who depicted the historic and current economic trends; Victor Sadras for his singular comments and careful revision of the draft; Huub van Lier, Rob Jongman and Ron van Lammeren from Wageningen University; Andrew Millington and Thomas Tanner from the University of Leicester; Ceferino Zárate, Serafina Cruz, Festo Chauque, David Sarapura and Jorge Pereda who taught them about the singular values of the communities of Finca Santiago and Tinkunaku.

The editors are finally grateful to their families for their continuous support and understanding.

1

Introduction: ecological networks and greenways

1.1 Nature conservation and landscape ecological principles

The concepts of ecological networks and greenways relate to the human environment and its development. Recent advances in environmental philosophy have shown the fragility of the concept of 'nature'. New avenues have been opened up, from the new genetic and reproductive technologies to the awareness that the continuous loss of nature is a 'foundational concept, a ground of being, a stable otherness to the human condition' (Robertson *et al.* 1996).

Yet the beginning of the third millennium reveals a moment in which changes in social, scientific and technological sectors are rapid and multiple. In a dynamic environment changes in nature too become more and more human-driven, while the landscape becomes human-dominated. The natural environment has thus been gradually fragmented and now retains sets of habitats and species that cannot survive in isolation.

Within this framework, new philosophical directions in environmental sciences have stressed the importance of moving from isolation to connection and from a concentric to a peripheral approach. Nature conservation, accordingly, is moving from local to global. If the previous focus was primarily on areas of high nature concentration, e.g. national parks, now the focus is moving towards linkages between them and linkages between nature and the human environment such as greenways, ecosystem coherence and ecological networks. These concepts have recently become familiar in ecological language at both the scientific and the public level.

The above considerations are significant for environmental conservation and sustainable development, which should in turn become priority issues for national and regional authorities. The notion of 'environment' comprises in its broadest sense all factors that are of importance for living species and living communities. Environmental conservation and management reach far

1

beyond technical environmental protection such as air and water purification. They also include maintaining the functioning of ecological systems in all their variety of spatial forms. Protective measures for separate environmental factors, such as emission reduction and noise protection, are not sufficient for a sustainable environment. An integrated approach based on landscape ecological principles is becoming ever more important.

The ability of humans to change landforms and land functions has made urgent the need for new avenues in planning decisions. Landscape ecological concepts have laid new foundations for basic planning ideas. In the past decades the integration of the multiple activities occurring on the land has been reconsidered in physical planning, impelled by the increasing academic interest in multidisciplinary approaches. This has important implications for the ecological functioning of the landscape.

Landscape ecological principles have been integrated into nature conservation and landscape planning. Site-based nature conservation can only be successful if the conservation sites are very large, as in Russia, but even then larger carnivores are threatened. Species have difficulty surviving in fragmented landscapes and this hampers the core objective of nature conservation. Consequently, nature conservation in Europe has changed from site protection to the conservation of ecological networks, including the wider landscape (Jongman 1995).

Plants and animals are dispersed by both wind and water, with the help of other species, or by their own movements. Migration is a special kind of dispersal, directed to a certain site; dispersal is essential in population survival and the functioning of biotopes. However, dispersal can only take place if there are means for dispersal and sites to disperse from and to. On the one hand, an animal species will leave a population if living conditions cannot support all its individuals; on the other hand, new species will fill the gaps in those sites or populations that become empty. As such, fluctuations in populations can cause changes in the species abundance and species composition of a site. Birth, death, immigration and emigration are the main processes that regulate fluctuations at the population level. Plants, like several other groups of organism, depend on other species for their dispersal. Nevertheless plant strategies for dispersal are the least known and most difficult to detect in practice, and in this book they do not play an important role. In general, restriction of species dispersal increases the chance of species extinction.

The main elements for dispersal in the landscape are the distance and the land use between sites, the presence of corridors, and the barrier effect of landscape (Opdam 1991). Area reduction causes a decrease of populations that can survive and an increasing need for species to disperse. Routes for species migration consist of zones that are accessible for the species to move from one site to another. Migration routes moreover are manifold, from single wooded banks to small-scale landscapes, and from river shores to whole rivers and coastlines. Migration, furthermore, is a prerequisite for many species to survive the winter period. However, migration is also risky and for flying animals migration routes should have as few barriers as possible and stepping-stones should be available for feeding, rest and shelter. For fish this means that rivers should not be blocked by dams and should have good water quality. For mammals and amphibians it means that routes should be available and that man-made barriers can be crossed.

1.2   Concepts and dualities in ecological networks and greenways

The linkages between ecology and landscape, as illustrated above, show the relevance of the discipline of landscape ecology to the development of ecological networks and their integration in the planning system. They also lead to another linkage, namely between ecology and networks.

The Concise Oxford Dictionary (1995) defines *ecology* as 'the branch of biology dealing with the relations of organisms to one another and to their physical surroundings'. It also defines a *network* as (a) 'an arrangement of intersecting horizontal and vertical lines, like the structure of a net' and (b) 'a group of people who exchange information, contacts, and experience for professional or social purposes'. Hence in certain instances the two meanings of 'network' have brought about confusion over what the word actually means, whether it is ecological or human oriented.

An *ecological network* is today recognised as a framework of ecological components, e.g. core areas, corridors and buffer zones, which provides the physical conditions necessary for ecosystems and species populations to survive in a human-dominated landscape. The goal should be considered twofold: to maintain biological and landscape diversity, but also to serve as a network assisting policy sectors in the conservation of natural ecosystems.

Again in *The Concise Oxford Dictionary* one definition of *green* reads 'concerned with or supporting protection of the environment as a political principle' and among the several definitions of *way* there is 'a place of passage; a course or route for reaching a place'. The blend of the two words produced *greenway* with an initial meaning of a passage for people and their access to the countryside, adding the function of linkage between the urban and rural American landscape (see President's Commission 1987).

A more comprehensive meaning has been proposed later by Ahern (1996) as networks of land planned, designed and managed for various purposes, but anyway compatible with sustainable land use. This definition incorporates a range of functions and possible typologies that can be set up within different contexts, e.g. natural, cultural, spatial or political. It is actually this diversity which is, as pointed out by Ahern in chapter 3, the tool for proper communication and the key to the popularity of greenways.

Ecological networks and greenways clearly show a distinction in approach and function, but similarity in concept and structure. While greenways came initially from the need to create connections and paths for people to access the American countryside, ecological networks came from the need to conserve European species and habitats. In their later stages, however, the two concepts have come closer, having both been recently recognised as fundamental frameworks for the survival and movement of species populations, including humans.

Greenways have certainly inspired the development of ecological corridors in European nature conservation at different levels. From the European Union directives (e.g. the Habitats and Species Directive with Natura 2000) and the Council of Europe strategies (e.g. the Pan-European Biological and Landscape Diversity Strategy with Action Theme 1) the greenway concept has influenced national and regional policies throughout Europe. But whereas an ecological corridor is just a component of the ecological network, a greenway constitutes the connectivity framework itself.

1.3 A new perspective in landscape ecology

Reviewing recent developments in ecological networks, Arts *et al.* (1995) concluded that 'during the last decade, the nature conservation policies in many European countries have been based on landscape-ecological research, especially concerning the role of land use and landscape structure in the survival of species and in the protection of nature reserves'. Landscape ecology indeed provided the insight that nature is at the landscape level a dynamic system reacting to a complex of environmental and land use conditions. Land use influences the functioning of ecosystems as a whole, its self-purification capacity and the carrying capacity of the landscape (Mander *et al.* 1988; Kavaliauskas 1995). It also affects habitat quality for wild species and the potential for dispersal and migration that are vital for survival of populations especially in fragmented landscapes.

In an ecological sense isolation is an important issue. It allowed speciation on islands and led Darwin to detect this phenomenon on the Galápagos and develop his basic evolution theory. Time made it possible for species to develop. In the recent much more dynamic world, isolation is an important feature of agricultural landscapes, for example in north-west Europe. Here processes do not play a role on the timescale of speciation but on that of population dynamics.

Isolation does not only occur in places where it is evident, but also in less obvious ecosystems such as plantation forests, where management can cause isolation of the remnants of natural old growth forests within them (Harris 1984). Most natural and semi-natural habitat sites are remnants of a former natural area. Present landscapes are dominated by man-made dynamic habitats, and the less dynamic habitats are small and isolated as are the populations in them. Habitat isolation and habitat loss prevent natural species from developing viable populations or let populations survive at different equilibrium levels (Hanski *et al.* 1985). Natural interconnections have declined with the disappearance of forested and river corridors, and with the development of human infrastructures. The strategy to overcome this is the redevelopment of ecological coherence through networks.

Ecological networks and greenways have clearly brought about a new vision of landscape ecology. Previous concepts have dealt with the single elements of the network, such as patches and nodes, buffer areas, corridors and linkages; or with the dynamics of the network, such as movements, flows, migration, dispersal, fragmentation and connectivity. We now deal with the entire framework, where the single elements interact with each other in a dynamic way, and are combined to create that ideal network for conserving biological and landscape diversity and at the same time supporting sustainable land use. Planning is the key that allows these goals to be reached, since creating ecological networks or greenways requires an established planning system, be it regional or local.

The protection of habitat and species, however, is affected by many factors interrelated with each other and with the human world. From the latter, variations to the environment can be triggered not only by planning, but also by administrative skills, community values, sense of place, environmental attitudes, political and economic situations. After all the final goal of ecological network and greenway development is nature conservation; the former is only a tool to reach the latter. Ecological networks, as suggested by Bouwma *et al.* in chapter 6, are one of the possible measures in tackling species and habitat conservation. A combination of several measures is hence the best practice for ensuring alternative solutions to the problem of environmental fragmentation. Moreover, since ecological networks consist of both ecological and human components, the interaction between nature and culture is a priority to consider in both nature conservation and sustainable development.

All this leads to the further perspective on landscape ecology envisaged in this book, namely the indispensable link between nature and culture in the implementation of future econets and greenways. In this new scenario the full ecosystems are the actors, nature the scenery and the human species the director. The last has great responsibility towards future generations and therefore has to accept a role that truly respects nature. From this standpoint one may see how econets and greenways have to be MULTI-functional, -cultural, -organisational, -national, being in turn part of the above-mentioned planning, social, economic, political and continental sectors. This book is an attempt to illustrate such a multi-sectorial dimension within the context of different countries.

2

The context and concept of ecological networks

2.1 Introduction

Niels Holgerson travelled on the back of his goose with the wild geese to the north to the breeding areas and he saw the land beneath him. Geese fly north and south for breeding and wintering. When walking in a forest or through agricultural land you may suddenly meet a deer, a badger, a hare or a rabbit. Looking carefully in the grasslands in the wintertime you can detect mouse tracks. Salmon migrate from the ocean up the rivers for spawning. Humans used to walk or drive horses to travel for business and religious purposes. In the modern western world, humans walk mainly to enjoy a tour on a grande randonnée or a old pilgrims' path. However, in general connection and exchange are important for all species to survive, to forage, to reproduce or establish new communities. On all scales and for most species pathways do exist to link individuals and populations.

Land use and nature changed dramatically in the nineteenth century. The land has been developed, parcels enlarged and land use intensified. Since the end of the nineteenth century nature conservation through national parks and nature reserves has been the prime tool to counteract decline of species and natural ecosystems. This was very important to preserve species and natural ecosystems. In that period the linkage between protected sites did not seem to be crucial. The increasing pressure of land use in combination with the insights from landscape ecology on the role of landscape flows and the functioning of metapopulations of species has changed that idea. Nowadays the idea is accepted that reserves alone cannot maintain biological diversity in the long term (Beier and Noss 1998; Bennett 1999). Networks and landscape linkages are needed for long-term survival.

This chapter introduces the reader to the development of concepts of ecological networks in general and the European approach to ecological networks in particular. In the last decades of the twentieth century ecological networks have been developed by authorities and scientific institutions in Europe, America and Australia (Bennett and Wit 2001). In Europe this has been partly a Europe-wide approach as a reaction to the Convention on Biological Diversity or as newly developed European policy, such as NATURA 2000 (Habitat and Species Directive, EC 92/34), the Emerald Network and the Pan-European Biological and Landscape Diversity Strategy (Council of Europe *et al.* 1996). These have been scientific studies and national initiatives varying from strategies for adaptation of conservation policies to development of ecological networks as a scientifically based approach. Nowadays most of the ecological networks in Europe are part of national and regional nature conservation policies.

This chapter does not intend to present a complete picture of these developments, but to present common denominators, concepts to highlight common developments between countries and regions, and indicate where and when differences between countries and regions have to be taken into account when developing ecological networks. This chapter emphasises the development in Europe and it shows trends in decline of landscapes and the diversity of approaches to biodiversity conservation and nature conservation planning. Understanding these differences and common issues is of the utmost importance to find common principles and approaches and to know when differences are important enough to be maintained or when they can be abolished.

2.2   A short nature conservation history

Throughout the centuries, land use was adapted to the restricted technical ability of people to change the land. In Europe this led to a rather stable pattern of landscapes until the second half of the nineteenth century. Then around 1850 the industrial revolution started. It meant a revolution not only in the urban environment, but also in the rural environment. Machines were introduced as well as fertilisers and wire fencing. Semi-natural areas were converted into agricultural land and the scale of agricultural holdings increased. In the same period people began to regulate the main European rivers, such as the Rhine, the Danube, the Elbe, the Meuse, the Po, the Rhone and the Tisza. That meant better transport facilities and a safer and better-drained land. This process started on a small scale, but has continued until the present day.

The history of nature conservation and of urban ecological networks started as a reaction to the industrial revolution. Already in the last half of the nineteenth century and the first period of the twentieth century nature was integrated into urban planning, for instance when the main axes of towns were developed into green boulevards, such as the Champs Elysées and the footpaths along the Seine in Paris (Searns 1995). In the USA Frederick Law Olmsted proposed in 1860 a plan for Brooklyn and later for Boston to link the urban parks and quarters by green corridors, the 'parkways'. These are routes to and from the urban parks surrounded by trees and with an aesthetic and recreational function. These parkways were between 65 and 150 m wide. The Boston plan is still partly extant and known as the 'Emerald Necklace', and also has a drainage function for the town. Drainage remained an important function of plans for green corridors through the whole century. In chapter 3 the historic developments in the USA are described in detail.

In England in the same period Ebenezer Howard (1898) developed the greenbelt concept, meant to regulate the urban sprawl of London and other urbanising areas in England by surrounding the inner city with parks about 8 km wide. Behind this belt commercial and industrial areas should develop. London is the clearest example of this approach. The difference between the two approaches is the linking function of the parkways and the dividing function of the green belts.

In other countries, such as the Netherlands, developments took place comparable to those in the USA and England. The industrial revolution had a heavy impact on the cities and a need for urban green developed: the Amsterdam Vondelpark was one of the first urban parks, established in 1870. The Dutch housing law of 1901 allowed town authorities to designate areas as open space for public use. The city of Arnhem was, around that period, the first town that officially designated a park as 'public green space' and there the park system still reaches from outside town into its city centre.

Nature conservation and urban development joined forces in the 1920s. At the international congress on housing and urban development in Amsterdam in 1924 the statement was made that nature is important for outdoor recreation, for its scenic beauty and its intrinsic value. Urban planners and architects pleaded for the development of parkways and in this period several were constructed, such as those in Rotterdam and Utrecht (van Langevelde, 1994).

After the Second World War nature conservation focused more on the preservation of values within semi-natural landscapes. This was especially important in the northern states of Europe, where the decline of nature was alarming. After the first nature conservation year, 1970, changes took place in nature conservation in western Europe; nature conservation acts were revisited in many western European states, in some cases by amending existing legislation and in other cases by formulating a new and more integrated nature conservation policy relating issues such as recreation, urbanisation, regional planning and agriculture. In this period nature conservation obtained a more or less accepted position in policy. The same process took place in central and eastern Europe after 1989; since then some central and eastern European countries have introduced the most progressive legislation based on the latest scientific knowledge.

2.3   The driving force: land use change

Under the influence of changes in food demand, caused by demographic trends, the cultivated area of Northern America and Europe has shown considerable fluctuations. Agricultural areas are moving from one region to another, forests are removed in one part of the world and forests of exotic species are planted elsewhere. Agricultural productivity in Canada, USA and the European Union (EU), measured in kilograms of dry matter per unit of area, continues to rise thanks to ongoing advancements in agronomic knowledge. Through the consequent changes in agriculture and forestry practices, landscapes have suffered rapid and often irreversible changes. These changes consist of two groups (Fry and Gustavsson 1996):

- changes resulting from the marginalisation of farmland and forests and consequent abandonment of earlier practices
- changes arising from the more intensive use of highly productive land, resulting in less land being farmed, but farming and forestry becoming more intensive, more specialised, and at larger scales

The result of these two opposite trends is a polarisation between intensively used land and natural or abandoned land. Intensifying agriculture makes land monofunctional and takes away both cultural and natural diversity. Intensification by one farmer – reducing production costs – will improve his position on the market. We also have to realise that the farming market is an international market. The farmers in the Paramó of the Andes have to compete with the large-scale potato farmers in Canada, and Greek farmers have to compete with Dutch and Danish farmers on the cheese market. Also the trade in animal stock is international or continent-wide. If the market is not regulated the farmers in the less favoured regions will marginalise and eventually abandon their land. Both intensive and extensive land use are expressed in the landscape: the structure of the land, the size of the parcels and the area of natural and semi-natural vegetation that is still present.

The pressure of economic competition in farming, forestry and urbanisation makes the land partly homogenised through the disappearance of regional differences in (semi-) natural features. This is not a new process but its features become more and more recognisable. We develop towards a homogenised world. As a consequence, the multi-functionality of the landscape is disappearing and outdoor recreation and nature conservation have fewer opportunities in the wider countryside. According to research by van Rabenswaaij et al. (1991) the optimum for the presence of critical meadow birds turned out to be the range between 50 and 150 kg N ha^{-1}. If this is applied to the practice of present day farming it appears that this regime can only lead to marginalisation (De Wit, 1992). Economically sustainable agriculture is being driven either towards an environmental optimum (lowest inputs and optimum output) or towards an economic optimum (high inputs and high outputs). Such land use can only be of significance to the lower valued vegetation types and non-critical meadow birds.

The results have been evident for decades in the European landscape. We see the decline in historical structural diversity of the landscape and that makes landscape coherence disappear. In the Netherlands, forests in the floodplains decreased from 1900 to 1980 by 90% and hedgerows by 80% (Jongman and Leemans 1982). In the period 1950–1990 all open side channels along the major branch of the Rhine disappeared (Jongman 1992). In the period 1976–1986 the tree-lines in the agricultural landscape of St Oedenrode, the Netherlands, decreased by 35% (without the roadside plantings even by 45%). In Provence, France, the disappearance of the silvo-pastoral system led to a simplification of the landscape structure (Hubert 1991). In Great Britain the number of plant species in infertile grasslands declined by 13% between 1978 and 1990, and in upland forests even by 20% (Bunce *et al.* 1999). In the Czech Republic the semi-natural elements disappeared nearly completely from the cultural landscape during the collectivisation process (Figure 2.1).

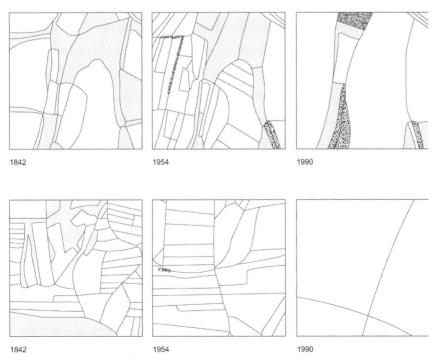

FIGURE 2.1.
Structural changes in the Czech rural landscape between 1842 and 1990 in two sample 1-km squares (Lipsky 1992). The collectivisation process took place between 1954 and 1970 and reduced the structural diversity of the landscape dramatically.

2.4   New functions in the landscape: fragmentation

Polarisation, homogenisation and marginalisation are not the only processes ongoing in the European landscape. In western Europe, intensive agriculture used to be an important land use in the urban fringe. Now its role is strongly diminishing, other urban-based functions take over, such as horse-keeping, garden centres and recreation facilities (Lucas and van Oort 1993). This trend is comparable elsewhere in Europe, from Lisbon to Moscow. In competition with urban functions, most rural functions cannot survive. The capital available in the cities will buy new functions in the surrounding rural landscape. This leads to more intensive use, more fragmentation of the landscape due to building, fencing and roads, and less space for natural elements. In addition, outside the urban fringe increasing road density, the building of new railroads and the intensity of land use lead to an increase of barriers in the landscape. The landscape develops into a new diversity of artificial elements causing fragmentation of natural features that can be considered as negative landscape diversity. Many animal species are sensitive to this kind of fragmentation. The area that they need for living depends on their home range and the maximum dispersal area. For small species roads are often inaccessible barriers. Some animals, such as amphibians in spring, take the risk of crossing roads towards breeding ponds. They are only successful in areas with low-density traffic. Larger animals will be hampered in their movements by urban areas, roads and unattractive land.

Urbanisation, agriculture and industry have put increasing pressure on the total area of landscape and nature. It is not only the claim for space that is important, but also the fragmentation of the landscape and the natural areas, especially in north-west Europe. This process of fragmentation has resulted in loss of habitats, fauna casualties, barrier effects, disturbance (noise and light) and local pollution (IENE 1997). Transport infrastructure in Europe (roads, waterways and railways) intersects living areas of populations of different species and in this way decreases the ability of the populations involved to disperse between different parts of their habitat or move to potential new habitats.

Fragmentation is caused not only by barriers, but also by a decrease in landscape elements (small forests, hedgerows, riparian zones). Fragmentation of natural areas is a spatial problem that can be defined as the dissection of the habitat of a species in a series of spatially separated fragments (Figure 2.2). It leads to a diminishing habitat area and spatial discontinuity.

Ecological effects are species-specific and depend on the size of the functional area, the mobility of the species and the isolating effects of artefacts in the landscape (roads, urban areas, canals). Both isolation and decrease of functional area of a habitat site increase the chance of local extinction of populations and

FIGURE 2.2
Fragmentation of the
landscape seen by an
imaginary species
(Opdam 1991).

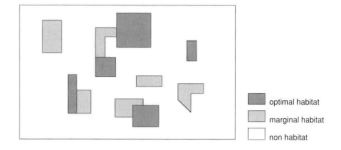

optimal habitat

marginal habitat

non habitat

diminish the chance of the spontaneous return of species. The spatial effects
are (Mabelis 1990):

- decrease in suitable area of the original ecotope
- increase in landscape heterogeneity and land use
- landscape fragments with subpopulations
- source–sink relationships in natural populations (larger natural areas
  become increasingly important for the survival of populations)

2.5   Landscape ecology concepts in landscape planning

2.5.1 Environmental conservation

Environmental conservation is a phrase that is used for a complex activity dealing with nature conservation and sustainable management. Environmental conservation and sustainable development of natural resources
should be priority issues for national and regional authorities. Environmental conservation and management reach far beyond technical measures for
environmental protection such as air and water purification. They also include maintaining the functioning of ecological systems in all their variety
and forms. An integrated approach based on landscape ecological principles is
becoming ever more important.

The land cover pattern of Europe consists of large areas of cultural landscapes, made by people and expressing the diversity of the regional climate and
soils and relatively small natural areas. Because of its interwoven position the
structure change of the European landscapes has been one of the major factors
in the deterioration of nature. Protection by law or by ownership cannot prevent species decline; there are too many relationships through water, air or
species movements. This means that spatial relations have to be included in nature conservation in the same way as for other human activities. Spatial planning is therefore indispensable.

The objective of spatial planning is to organise functions and space in such
a way that it shows the best mutual relationship, or to develop human and

natural potentials in a spatial framework in such a way that all and everything can develop as well as possible (Buchwald and Engelhardt 1980). What has characterised the concept of spatial planning in Europe, at least since the 1920s, has been the institutionalisation of planning concepts such as segregation of functions and functionalism. Planning is rooted in changes in social life: growing urbanisation and the territorial demands of an increasing population lead to separation of living and working space. In spatial planning the ideas of segregation have been pushed towards the extreme edge in the human exploitation of nature. Land use with its impact on the environment and nature has urged the development of more integrated approaches to planning that include landscape ecological concepts. In the past few decades, integration of functions has been restored into favour in planning and an increasing interest in multi-disciplinary approaches can be observed.

Landscape ecological principles have become part of nature conservation and landscape planning. Site-based nature conservation can only be successful if the conservation sites are huge, as in Russia. But even there larger carnivores are threatened. Species have difficulty in surviving in fragmented landscapes and the conflict between nature and society hampers nature conservation. The new way forward is now that nature conservation in Europe is changing from site protection into conservation of ecological networks, including the wider landscape (Jongman 1995). That means that nature conservation principles have to be included in spatial planning, as is already the case in several countries such as Denmark (Jongman and Kristiansen 2001) and the European Spatial Development Perspective (European Commission 1999).

Three landscape ecological concepts (ecostabilisation, connectivity or connectedness and the spiralling concept) all help to understand separate aspects of landscape functioning that should be integrated in landscape planning. Together they form a strong landscape ecological basis for landscape planning.

2.5.2   Ecostabilisation

The German and Eastern European tradition in applied geography has concentrated on regional relationships and has found applications in physical planning. Spatial planning in the Soviet era was subordinate to the rules of the planned economy. This kind of planning initiated large-scale technocratic projects and a mono-functional simplification of the collectivised agricultural landscape. Until the end of the 1980s nature did not figure in the spatial planning maps of countries such as East Germany (GDR), and zoning was directed by economic principles. Destruction of traditional landscape systems and ecological destabilisation of landscapes, with extensive erosion, salinisation of

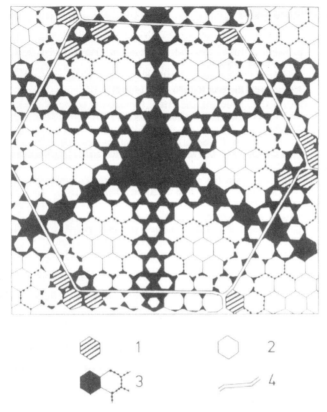

FIGURE 2.3.
The ideal structure of the landscape according to the principle of the polarised landscape, consisting of ecotones and a network of compensative areas: 1, urban areas; 2, fields; 3, compensative areas; 4, main roads (Mander *et al.* 1988).

soils, forest die-back and water and soil pollution, were visible results of this process.

For mitigation of the impacts of economic planning in central and eastern Europe, the ecostabilisation principle was developed in the early 1980s, resulting in planning concepts such as '*territorial systems supporting landscape ecological stability*'. In theory the approach was based on the idea of a *polarised landscape* (Figure 2.3) that was worked out in 1974 by the Russian geographer Rodoman (Mander *et al.* 1995).

This concept accepts intensive land use, but proposes a functional zoning of the landscape, including areas and elements for natural zones as a contrast to the centres of intensive land use. These natural zones and the zones selected for agriculture, industry and urban development were the poles of this planning concept. Rodoman's concept was dialectic and holistic, developed within a deductive scientific tradition. His formal and geometrical principles were developed into principles for practical ecological planning in the Eastern and

Central European states in the late 1970s and the beginning of the 1980s and it was based on co-operation between geographers, ecologists and territorial planners (Kavaliauskas 1995; Buček *et al.* 1996; Miklós 1996). It resulted in concepts such as *'nature frame'*, *'natural backbone'*, *'ecological compensative areas'* and *'ecostabilising functions'*. Essential to these concepts are:

- the designation of territories to function as an ecological compensation to the territories that are heavily exploited
- the linkage of these compensative territories by zones with coherent land management
- the availability of sufficient space to create compensation and linkages

Because Rodoman's approach has been published only in the Russian language, the theory of a polarised landscape is only known in the West through its applications in other countries. In the western European countries a polarisation of the landscape and land use has become reality as well, and comparable planning concepts such as the framework concept in the Netherlands have been developed (Vrijlandt and Kerkstra 1994). This was done to optimise the consequences of the ongoing segregation of land uses. The competition for land, not least between agriculture and nature conservation, in the more densely populated areas in Western Europe has had a decisive influence on this development and has caused the debate about whether long-term ecological goals could be achieved with separation of land uses or whether a multi-purpose development should be the ideal of landscape planning in Europe.

The important principle behind the ecostabilisation concept for spatial planning is the acknowledgement of the importance of processes on the landscape scale, the presence of flows, the role of ecotones, and the use of the ability of nature to purify and restore.

2.5.3 Dispersal and migration, connectivity and connectedness

Movement is itself the product of evolutionary pressures contributing in many ways to the survival and the reproduction of animal species. Animals move through their home range, but may also move long distances from where they were born and their kin remain. Three kinds of movement can be distinguished (Caughley and Sinclair 1994):

- local movements: these are movements within a home range and are on smaller scales
- dispersal: movement from the place of birth to the site of reproduction, often away from its family group and usually without return to place of birth

- migration: movement back and forth on a regular basis, usually seasonally, e.g. from summer range to winter range to summer range

Local movements are within the home range of a species. They are movements for foraging, hiding from enemies, and optimising living conditions. Ricklefs (1990) defines dispersal as the one-way permanent movement away from established home range or natal area. In contrast, migration is the two-way movement between two areas. Dispersal is often associated with what is referred to as emigration and immigration, i.e. movement from one population to another. It is usually one-way movement (Caughley and Sinclair 1994). Stenseth and Lidicker (1992) claim that dispersal or movement from one home site to another is a phenomenon of potentially great importance to the demographic and evolutionary dynamics of populations. However, it is also one of the less understood features of ecology and evolution. It determines the probability that an individual currently 'here' will later be 'there', and as a consequence be exposed to different opportunities and risks. It gives populations, communities, and ecosystems their characteristic texture in space and time (Clobert *et al.* 2001). In general dispersal can have advantages and disadvantages for a species (Table 2.1).

Migration can be defined as a population's periodic or seasonal movement, typically of relatively long distance, from one area or climate to another. Migration is characterised by the periodicity and regularity of the movement. The term 'migration' is generally reserved for mass directional movements of individuals of a species from one locality to another. The individuals engaged in migration mostly move between two locations and return to the same home ranges and often with the same mates.

Why species migrate is still one of the most challenging questions, especially in ornithology (Berthold 1993; Pettingill 1970). According to Gill (1995) the potential benefits of migration are species- or population-specific, such as the need to escape from inhospitable climates, probable starvation, social dominance, shortage of nests or roost sites, or competition for food. For birds the advantage of migration is that migrants can exploit seasonal feeding opportunities, while living in climates that are favourable throughout the year. The benefit must be substantial, because the costs of migration are potentially large.

Migrating species are vulnerable in their life cycle. They are not available all year to signal the importance of a site as a temporary habitat. European storks (*Ciconia ciconia*), for instance, breed in central Europe and winter in Africa, migrating 10 000 km each season. The breeding population is mainly concentrated in Germany, Poland, the Czech Republic, Slovakia, Hungary and

Table 2.1. *Dispersal balance sheet*

Type of factor	Potential advantages	Potential disadvantages
Environmental	Escape from unfavourable conditions (economic, physical, social)	Uncertainties of finding food, shelter, appropriate social milieu
	Reduced exposure to predators or competitors	Greater exposure to predators or competitors
	Reduced exposure to population crashes	
	May find uninhabited or lower-density area of suitable quality	May not find any empty habitat
Genetic: quantitative	Promiscuity	Uncertainty of finding a mate
	Frequency-dependent selection may favour rare phenotypes	Strange phenotypes may be avoided
qualitative	Heterosis and avoidance of inbreeding	Fewer viable offspring may be produced (breaking down of co-adapted systems; disadvantageous recombination)
	Greater chance for new and advantageous recombinations occurring	

Source: Stenseth and Lidicker (1992).

the Baltic States in the east, and Spain and Portugal in the west. They used to cover a larger area, but their breeding success has been severely hampered by land use changes in the past few decades.

Species have adapted to the cultural landscapes of Europe, because they were accessible and not hostile. Large areas with good living conditions that are always inhabited are defined as core areas for populations. In good reproductive years species will move from these areas into other – even marginal – sites (Verboom *et al.* 1991). Area reduction will cause a reduction of the populations that can survive and in this way an increased risk of extinction, because dispersal between habitats decreases, causing less exchange of genetic information and less colonisation of empty habitats.

Plants and animals both disperse by wind, by water, with the help of other species, or by their own movements. Migration is a specific kind of dispersal, it is directed to a certain site. Dispersal is essential to population survival and the functioning of biotopes. However, dispersal can only function if there are sites to disperse from and to and means for dispersal. On the one hand, animal species will leave a population if living conditions cannot support all individuals; on the other hand, species will fill in gaps in populations or sites that became empty. Fluctuations in populations can cause changes in species abundance and species composition of a site. Birth, death, immigration and emigration are the main processes that regulate fluctuations at the population level. Plants, but also several other groups of species, often depend on other species for their dispersal. However, plant strategies for dispersal are the least known and most difficult to detect in practice. Restriction of species dispersal increases the chance of species extinction.

The main functional aspect in the landscape of importance for dispersal and persistence of populations is connectivity and connectedness. According to Baudry and Merriam (1988) connectivity is a parameter of landscape function, which measures the processes by which subpopulations of organisms are interconnected into a functional demographic unit. Connectedness refers to the structural links between elements of the spatial structure of a landscape and can be described from mappable elements.

Structural elements are different from functional parameters. For some species connectivity is measured in the distance between sites, for other species the structure of the landscape, such as the connectedness through hedgerows, represents the presence of corridors and barriers. Routes for species migration consist of zones that are accessible for the species to move from one site to another and back. Owing to differences in needs, migration and dispersal routes can be manifold, from single wooded banks to small-scale landscapes and from river shores to whole rivers and coastlines. Migration is a prerequisite for many species from northern Europe to survive the winter period. Migration is risky; for flying animals, migration routes should have as few barriers as possible and stepping stones should be available for feeding, rest and shelter. For fish, rivers should not be blocked by dams and should have good water quality. For mammals and amphibians, routes should be available and man-made barriers should be able to be crossed.

Amphibians and mammals are able to disperse over distances from several metres to hundreds of kilometres. For small mammals ecological corridors can be hedgerows, brooks and all kinds of other natural features that offer shelter. Migration is important for grazing animals such as red deer (*Cervus elaphus*) and roe deer (*Capreolus capreolus*), for predators such as the golden eagle (*Aquila chrysaetos*), the lynx (*Lynx lynx* and *L. pardina*) and the wolf (*Canis lupus*), and also for most birds from northern and eastern Europe.

The importance of metapopulation principles in planning is the acknowledgement that survival of species is more than solely maintaining nature reserves; ecological linkages are needed and must be included in spatial plans (Figure 2.4).

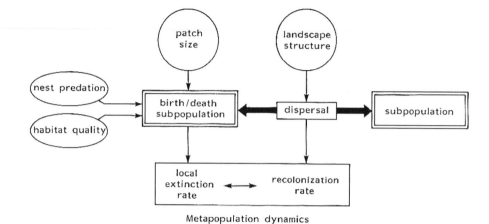

Metapopulation dynamics

FIGURE 2.4.
Factors influencing metapopulation dynamics according to Opdam (1991).
Dispersal is a functional aspect of the different metapopulations. How dispersal is
realised depends on the requirements of the species and the landscape structure
present.

2.5.4 Rivers: the continuum, flood pulse and spiralling concept

A river is more than the sum of its parts and it is not a static body of water, but rather a continuum with a changing ecological structure and function. The concept dominating river studies over the past few decades was the River Continuum Concept, which was the first unified hypothesis about how streams and their watersheds work. The River Continuum Concept (Vannote *et al.* 1980) is based on macro-invertebrates and it states that from the headwaters to the river mouth a continuous change in macro-invertebrate community takes place. The headwaters are likely to be narrow, fast-flowing, and shaded by trees and other vegetation, therefore virtually all the energy that enters is in the form of leaves, twigs and other debris. Detritivores and filter-feeders dominate the fauna. Energy in the form of biomass and detritus is constantly flowing downstream, hence the energetics of any particular section of the river are influenced by events upstream. The result is a longitudinal continuum of ecosystem structure with a number of predictable properties. Vannote *et al.* (1980) argue that the river's biological and chemical processes correspond to its physical attributes and that the nature of biological communities changes in a downstream direction just as the river itself does.

Running waters are, however, far more than just longitudinal river corridors and according to Jungwirth *et al.* (1998) modern ecology recognises them as complex ecosystems. According to Townsend and Riley (1999) the science of river ecology has reached a stage where explanations for patterns rely on links at a variety of spatial and temporal scales, both within the river and between the river and its landscape. The links according to Townsend and Riley (1999) operate in three spatial dimensions:

- longitudinal links along the length of the river system, such as the river continuum (Vannote *et al.* 1980), with downstream barriers to migration
- lateral links with the adjacent terrestrial system, such as the flood pulse concept (Junk *et al.* 1989)
- vertical links with and through the riverbed

Many linkages occur between the river and its environment, therefore it is suggested that the river continuum be considered within broad spatial and temporal scales (Roux *et al.* 1989). The lateral and vertical dimensions of the ecosystems need to be associated with running water. In the flood pulse concept (Junk *et al.* 1989) it is stated that the pulsing of the river discharge, the flood pulse that extends the river onto the floodplain, is the major force controlling biota in rivers with floodplains. The flood pulses control biota in three ways: directly by facilitating migration of animals, indirectly by enhancing primary production, and thirdly by habitat structuring. During floods biota migrate both actively and passively between different habitats in the river floodplain system. The lateral exchanges between main channel and floodplain, and nutrient recycling within the floodplain, have, according to Grift (2001), more direct impact on biota than the nutrient spiralling discussed in the River Continuum Concept. The floodplains also provide important factors for driving ecological processes in the riverine ecosystem, mainly the habitat complexity and habitat quality in the river ecosystem.

The distinction between dispersal from the breeding quarters over a wide stretch of territory, and the more clearly defined migration, may be useful when applied to bird migration (Jones 1968). However, fish may be carried passively hundreds of kilometres by currents and what may be no more than dispersal could then have all the appearance of a true migration. Northcote (1978) suggests that also fish migration has evolved in a way comparable to that of mammal and bird movements, to optimise feeding, to avoid unfavourable conditions, to enhance reproductive success and to promote colonisation of new habitats. Depending on the environmental situation and species characteristics, fish can be considered to be dispersing or migrating. Fish migrations in rivers are associated with currents, although during the life cycle the direction of fish movement with respect to the current often changes. Active migrations against the current (spawning migrations) generally occur together with passive or active–passive, downstream migrations of juveniles and recently spawned brood stock. In some species individuals are capable of migrating long distances and returning to their starting points accurately. The accuracy of return migration is widely variable but may involve a return to the actual place of birth, known in ichthyology as homing. Homing was defined as the return to a place formerly occupied instead of going to other, equally probable places suitable for reproduction, at the time when other sexually mature fish are also present (McDowall 1988).

There are more types of fish migration. Bemis and Kynard (1997) distinguish between one-step and two-step migrations. Defining the one-step spawning migrations as those in which fish move directly upstream with the dependence on the bio-energetic reserves of the fish, the migration may be short or long and occur in winter or spring. Two-step spawning migrations involve upstream migration, usually in the autumn, followed by overwintering near the spawning site, followed by a short or long migration to spawn the following spring. Besides the longitudinal migration the fish migrate between the main channel and the floodplains, utilising the floodplains either temporarily or permanently during their life cycle. The backwaters and inundation areas are important spawning, foraging and nursery places, or places used to avoid harsh conditions (Jurajda 1995).

In general, running waters constitute a vector for the transfer of material from elevated reaches to the bottom of a drainage basin. Fish, mammals and plants move along their corridor at different speeds and with different steps. The strong interaction between the stream and its riparian ecosystems in its ecotone provides a huge exchange of energy, matter and nutrients that attracts all kinds of natural species. The transport of matter and nutrients is restrained by all kinds of natural and man-made retention devices and in this way the river is an important mechanism for reconstruction of landscapes and for species, linking reproduction sites and populations. The way in which matter, energy and species move through a river system can be well described by the spiralling concept, based on the recurrent use of matter in ecosystems along the river (Figure 2.5). Although this concept has been developed for nutrients, it can also be a valuable concept for considering the behaviour of species.

Rivers are in principle dynamic systems and in relation to other parts of the landscape important pathways. Downstream they have higher diversity and are less dynamic than upstream. This also means that interactions with other ecosystems are more complex in the downstream stretch than upstream. Human use of rivers as a transport route or recreation pathway will also be different, because of differences in accessibility, the diversity of the landscape and the potential use of the system. It varies from water provider to the most intensively inhabited areas in the world, where conflicts for space are common.

Rivers are mostly not considered as one system. In general, several authorities and users decide on maintenance and use, because of the many borders that are crossed and the many interests that are involved. Ecology of rivers

FIGURE 2.5
The spiralling concept for river systems
(Pinay *et al*. 1990). Strong interactions
between the stream and the riparian
systems in the ecotone represent important
exchanges of nutrients, matter and species.

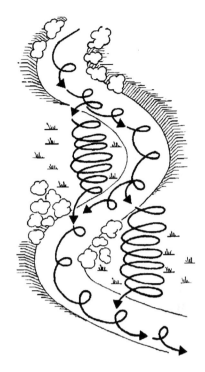

is a difficult and little-known aspect. Transport of waste and fish migration only become evident if large accidents occur such as the Sandoz accident in the 1980s, or the Guadiamar and Tisza accidents in the 1990s. These emphasise the role of rivers as ecological corridors.

Owing to their role as transport mechanisms for nutrients, matter and species, rivers should be key systems in spatial planning and at least they are in the development of ecological networks. The spiralling concept gives a good understanding of the lateral and longitudinal processes in rivers and the way they can be anticipated in planning. For a further elaboration of the planning of the river corridor of the Guadiamar, see chapter 15.

2.6   Ecological networks

2.6.1 Structure

Landscape ecology provides the insight that at the landscape level nature is a relatively dynamic system reacting to a complex of environmental and land use conditions. Land use is considered to influence the functioning of an ecosystem as a whole, its self-purification capacity, and the carrying capacity of the landscape (Mander *et al*. 1988; Kavaliauskas 1995). It also affects habitat quality for wild species and the potential for dispersal and migration that are vital for survival of populations, especially in fragmented landscapes.

In ecological sense isolation is an important feature in agricultural land-scapes of northwest Europe. Even in plantation forests, management can cause isolation of the remnants of natural old growth forests within them (Harris 1984). Most natural and semi-natural habitat sites are remnants of a former natural area and the species in it are part of a metapopulation that has sur-vived. Present landscapes are dominated by man-made dynamic habitats and the less dynamic habitats are small and isolated, as are the populations in them. Habitat isolation and habitat loss prevent natural species from developing vi-able populations, or let populations survive on different equilibrium levels (Hanski *et al.* 1985). Natural relations have declined through the disappearance of forested corridors and natural river corridors and the development of human infrastructure. The strategy to overcome this is the redevelopment of ecological coherence through networks.

Ecological networks can be defined as systems of nature reserves and their interconnections that make a fragmented natural system coherent, so as to support more biological diversity than in its non-connected form. An ecolog-ical network is composed of core areas, (usually protected by) buffer zones and (connected through) ecological corridors (Bischoff and Jongman 1993). Core areas have mostly been identified by traditional nature conservation poli-cies. The insight gained from recent geographical and ecological concepts link this traditional conservation strategy with other land use and integrate na-ture conservation into general land use policy and spatial planning. In this way ecological corridors and buffer zones are becoming key elements in nature conservation strategy.

However, the existing protection system and the development of ecologi-cal networks as a new concept in a region or country are not just an exten-sion of each other. New concepts have to be applied and the history of nature conservation should be evaluated to detect gaps in the protection system that have to be solved and improve connectivity.

2.6.2   Core areas

In Europe many, but not all, important natural areas are protected. The definition of protected areas differs much between countries and the le-gal and property rights are regulated differently. Various attempts have been made to inventory protected areas by using a classification systems. The IUCN classification (IUCN-CNPPA 1982) of protected areas is generally accepted. To understand the protection status the national legal definitions should be taken into account. In Box 2.1 some of the definitions for national parks are pre-sented. Some countries such as Denmark do not have national parks. However, they have long-established legal protection of small biotopes (Brandt 1995).

BOX 2.1. Definition of National Parks in Europe

IUCN:
A relatively large area, where

- one or more ecosystems have not been changed fundamentally by human exploitation and habitation, where plant and animal species, geomorphologic objects and biotopes of special value occur or that contains a natural landscape of great beauty;
- the highest authority in charge of the country took steps to avoid potential exploitation as soon as possible, reduce settlement in the whole area and stimulate effectively the conservation of ecological, geomorphologic and aesthetic characteristics, that led to the initiative of its foundation; and
- it is permitted to visit the area under special conditions for the inspiring educational, cultural and natural values.

The area must be managed as a whole.

Germany:
A National Park is an area of a large size, little influenced by people, that deserves special protection because of its natural beauty and special ecosystems and where the core area is managed as a nature reserve.

The Netherlands:
A National Park is a single area of at least 1000 ha consisting of natural systems such as waters and forests with special conditions and plant and animal life. Good possibilities exist for zoning and recreational use as well. In a National Park hardly any agricultural land is found.

Great Britain:
A National Park is a large area mainly founded because of its great landscape and scenery values. Human settlement and human activities are usually present. The importance of the natural environment varies with the park; if present, natural values are situated in nature reserves in the park.

Greece:
A National Park is an area that is mainly forested and needs special protection because of

- flora, fauna, geomorphology, soil, air, water and natural environment in general;

- the necessity to keep the natural condition undisturbed or to improve it because of aesthetic values, spiritual welfare of people, or scientific research.

Italy:
A National Park is a large area:

- that is protected because of the presence of valuable flora and fauna, important geological formations and landscape beauty;
- that aims at the enhancement of recreation and tourism;
- that gives space to human exploitation to provide an income for local people; and
- where it is forbidden to hunt.

Portugal:
A National Park is a large area that can be found in remote parts of the country where people have managed the environment in the same traditional way for centuries. A National Park contains special landscapes and an important flora and fauna.

France:
A National Park is an area that is nearly uninhabited, with strict rules for conservation of flora and fauna, biotopes and special landscapes for visitors to the Park, and that is surrounded by a buffer zone in which tourist activities and the rural economy will be stimulated.

Differences in definitions used by countries in Europe can be big. Agriculture, forestry and recreation are allowed in some cases, in other cases an integral part and in another group not part of the protected area. Traditional land use, especially extensive exploitation of grassland, can be a method of management of semi-natural areas. Other protected areas are areas for landscape conservation, nature parks, areas of outstanding natural beauty, etc. These areas can include protected areas for nature conservation. Agriculture, forestry and recreation are more or less limited by rules concerning land use, buildings and environmental protection. Public access is regulated differently.

The national strategies for nature conservation in Europe have entirely been determined by national policies. Now, through the EU Habitats and Species Directive (92/43/EEC), coherence will be brought into the developments in Europe. However, national differences will continue to exist and be taken into account when designing and implementing ecological networks. This means that the definitions are also determined by national traditions and policy (Box 2.2).

BOX 2.2. Definitions of ecological corridors in European countries

Belgium, Flanders:
Nature corridor zones are zones that, without regard to their areas, are of importance for the migration of plants and animals between areas of the Flemish Ecological Network and/or nature reserves and which are strips or linear-shaped with a series of small landscape elements (De Blust, pers. comm., 1999).

Czech Republic:
Biocorridors are ecologically significant landscape segments that connect biocentres and support the migration, spreading out of and contact between organisms. Biocorridors can be separated into several groups according to state, function and degree of biodiversity. A *continuous corridor* contains communities with a high degree of biodiversity throughout its length. An *interrupted corridor* is broken by one or more barriers of human origin such as a road, a field or a built-up area. A *modal corridor* connects biocentres containing the same or similar communities. *Contrast biocorridors* connect biocentres containing distinctly different communities (Veronica 1996).

Estonia:
An 'official definition' (in the legal acts) of 'ecological corridor' is missing. Three main aspects of ecological corridors are usually emphasised: (1) an arrangement of habitats that enhances the movement of animals or the continuity of ecological processes through the landscape, (2) a general term for a linkage that increases connectivity at a landscape or regional scale, (3) a linear strip of vegetation that provides a continuous pathway between habitats.

Hungary:
Ecological corridor means any ecological passage made up of natural and semi-natural areas and strips, which ensure or support the ecological connection between distant territories.

The Netherlands:
Ecological corridors are areas or structures that enable spreading, migration and exchange of species between core areas and nature development areas of the National Ecological Network.

2.6.3 Buffer zones

The concept of buffer zones is rather old (Wright and Thompson 1935). The literature offers a number of definitions, related to the approach used for their design within the framework of spatial planning and management. IUCN defines a buffer zone as a zone peripheral to a national park/reserve where restrictions are placed upon resource use or special development measures are undertaken to enhance the conservation value of the area (Oldfield 1988). The more socio-economic approach is expressed by the World Bank definition: a social agreement or contract between the protected area and the surrounding community, where size, position and type of buffer zone are defined by the conditions of this agreement. Within the framework of an ecological network strategy the definition of a buffer zone should integrate both landscape and functional attributes. A definition based on the ecological function(s) of the buffer zone should focus on its main management objectives (Miklós *et al.* 1995):

- protection, to protect from harmful human activities
- interaction, to sustain positive landscape interactions
- diffusion, to sustain natural and man-made flows in the landscape

Buffer zones aim at controlling human activities within the lands adjacent to a core protected area by promoting their sound management, thus decreasing the potential impacts and the probability of isolation. The presence of a local population is implicitly permitted within the buffer zone (otherwise the buffer zone would be a totally protected area). The current approach in buffer zone design tends to accept them as areas where a plan of land-use regulations is applied rather than as clearly defined areas that could have legal protection. Thus, the buffer zone is (or should be) designed (Jongman and Troumbis 1995) to:

- protect local traditional land use
- accomplish area requirements or shape irregularities of the core area
- set aside an area for manipulative research
- segregate core areas for nature conservation from other land use such as agriculture, recreation or tourism activities
- manage adverse effects by putting up a barrier for immediate protection
- locate developments that would have a negative effect on the core area

Landscape change outside the boundaries of a core area generally causes important biotic changes within it. By creating environmental gradients, buffer zones maintain landscape processes and elements around the natural remnant

to avoid abrupt changes. Common sense and practical experience make us realise that the fundamental role of protected areas would not be achieved if the
controlling of adjacent human activities were not accomplished. Buffer zones
may be viewed as a shield around the core area against the direct impact of
human activities. Human activities should be viewed both macroscopically as
changing land use and landscape patterns, and at their physical dimension
as disturbances generating a long series of abiotic and biotic fluctuations influencing species and communities under protection (Jongman and Troumbis
1995).

2.6.4 Ecological corridors

   Connectivity and connectedness come together in the concept of ecological corridors. Ecological corridors can be defined functionally to indicate connectivity and as physical structures to indicate connectedness. They can be defined as functional connections enabling dipersal and migration of species that
could be subject to local extinction (Bouwma *et al*. 2002). As physical structures
they also can be defined as various landscape structures, other than core areas,
in size and shape varying from wide to narrow and from meandering to straight
structures, which represent links that permeate the landscape, maintaining or
re-establishing natural connectivity (Jongman and Troumbis 1995).
   As physical structures that can be identified within an ecological network,
they are mostly multi-functional landscape structures. In Europe ecological
corridors are often the result of human intervention in nature: hedgerows,
stone walls, landscapes with small forests, canals and rivers. Others, such
as coastlines and watercourses, are predominantly natural. The nature of
ecological corridors and their efficiency in interconnecting remnants and in
permeating the landscape depend on the habitat site they originate from and
the land use mosaic within which they are embedded and of which they consist.
Their density and spatial arrangement change according to the type of land use.
Their connectivity function varies from high to low depending on their spatial
arrangement, internal structure and management.
   Ecological corridors are multi-functional by definition. They have functions
for:

- *aesthetics*: they give an area its characteristic identity
- *social and psychological well-being*: they make an attractive living
  environment
- *education*: they help people to understand and experience nature
- *recreation*: nature close to housing
- *ecology*: temporary and permanent habitat and pathways for species

Ecological corridors are multi-functional in both an ecological and a societal sense, because they are not the core areas of a nature conservation system but function in the wider landscape. They are also part of greenways (see chapter 3). They can be as wide as a watershed or as narrow as a trail. They can encompass natural landscape features as well as a variety of human landscape features and are classified, from more natural to more cultural, as (Florida Greenways Commission 1994):

- landscape linkages, large linear protected areas between large ecosystems including undisturbed rivers
- conservation corridors, less protected and in many cases with recreational functions, often along rivers
- green belts, protected natural lands surrounding cities to balance urban and suburban growth
- recreational corridors, linear open spaces with intensive recreational use
- scenic corridors, primarily protected for their scenic quality
- utilitarian corridors, e.g. canals and power lines, that have an utilitarian function but also serve natural and recreational functions
- trails, designated routes for hikers and outdoor recreation, that have a function as natural corridors as well

This classification shows clearly the potential multi-functionality and morphological diversity of greenways and ecological corridors. The more complex a corridor is, the better it can function for different species groups and the more it is multi-functional in an ecological sense. A high immigration rate can help to maintain species number, increase metapopulation size, prevent inbreeding, and encourage the retention of genetic variation, and this can be judged as the main advantage of corridors (Simberloff and Cox 1987). They increase the foraging area for wide-ranging species and provide opportunities to escape predators and disturbances.

The ecological functions of corridors can be several:

- *dispersal*: range extension or redistribution of a population
- *migration*: to escape adverse environmental conditions
- *foraging*: movements between/nesting and feeding grounds
- *reproduction*: movement between winter sites and reproduction sites

A typology of ecological corridors can also based on their structure, shape and position in the landscape. As an example, in Slovakia the following types have been defined (Miklós 1996):

- According to their relative spatial position to core areas (biocentres):
  conjunctive corridor, connecting two core areas
  'blind' corridor, no core area at one end (peninsular wedging)
- According to their structure:
  continuous corridors, without gaps
  interrupted corridors, 'stepping stones', 'diffusion by jumps'
- According to their topographic position:
  on ridge positions, divides of watersheds
  in valleys
  on slopes (transversal)
- According to their shapes:
  line-like (typical example: ecotones)
  belt-like
  belt-like for water flows (a specific type of belt corridor)
  diffuse (created by a mosaic of different landscape elements without
    marked direction)

It must be stated that corridors can also have negative influences, such as the breaking of isolation that is needed for some species, exposing populations to more competitive species, the possibility of spreading diseases, exotic species, and weeds, disrupting local adaptations, facilitating spread of fire and abiotic disturbances, and disruption of local adaptations (Noss 1987b). Beier and Noss (1998) stipulate, based on empirical research, that ecological corridors for connectivity to maintain biodiversity are valuable conservation tools. Not maintaining or re-establishing ecological corridors (Figure 2.6) would mean that humans neglect the last remnants of natural connectivity and in this way could harm their own nature conservation objectives (Beier and Noss 1998). Moreover, modern experience shows that transport by people is much more important for spreading species and diseases.

2.6.5 Barriers

Planning of ecological corridors is a method for compensating for a long-term fragmentation process in agricultural landscapes. A network can meet all kind of barriers. Natural barriers exist at all levels. The Atlantic Ocean is a barrier between America and Europe for most plant and animal species. Mountains and rivers can be barriers for mammals and agricultural roads can already be barriers for insects and spiders (Mader 1988).

Increasing traffic and intensifying agriculture made the European cultural landscape more open on the one hand and more difficult to access on the other.

FIGURE 2.6.
Construction of a biocorridor in Moravia, Czech Republic (1992).

Hedgerows disappeared in intensively used agricultural land, forests became uniform plantations, streams have been straightened and the road network became asphalted, denser and more intensively used. Last but not least, many large and important wetlands have been drained. Canalisation of waterways and the building of motorways disturbed both the habitat of species and their ability to disperse.

Fish ladders (Figure 2.7) have to be built to make it possible for fish to cross weirs and locks. Road crossings can be made as tunnels or overpasses. Tunnels are used by small species. Habitat elements must be replaced at the correct side of the road and they have to be constructed in such a way that wild species are guided towards the tunnel. Overpasses or ecoducts are meant for larger species. Further explanation of the barrier role of infrastructure is given in chapter 5. A proposal to reduce traffic impact by traffic calming concepts is described in chapter 10.

2.7   Conclusions

The development of ecological networks in Europe is a logical development in the history of nature conservation. An increased understanding of the functioning of ecosystems and populations on the one hand, and the huge changes in the structure of the European landscape on the other, made scientists, policy makers and planners aware that traditional nature conservation

FIGURE 2.7.
Fish ladder in one of the headwaters of the Tweed (Scotland) to enable Atlantic salmon to migrate towards spawning grounds after decades of blocking by stream regulation.

through site protection would not be sustainable in the long term. Coherence between nature conservation sites is needed to let populations of several species survive. Ecological networks and greenways are an answer to that. The concepts and practical solutions under development are showing increasing coherence.

3

Greenways in the USA: theory, trends and prospects

3.1 Introduction

Development in Europe and the USA have common roots, but to a great extent they went their own way owing to geographical, political and scientific differences. Whereas ecological networks are common in Europe, greenways are much more an American product.

This chapter is organised to address several objectives:

- to compare classic and emerging definitions of greenways
- to articulate greenway theory, and to link definitions with theoretical principles
- to review the origin and evolution of greenways in the USA
- to discuss future prospects and research needs for greenways

Greenways is a 'new' word with many meanings. Much confusion still exists around its definition, yet it continues to gain in popularity and to appear regularly in popular language and planning policy in the USA and internationally (Fabos and Ahern 1995). The many differing perspectives on greenways are reflected in these definitions and serve to emphasise the complexity of the greenway concept. A brief review of these definitions provides a useful introduction to the subject of greenways, and underscores the need for a common definition and taxonomy, to support international, interdisciplinary communication and collaboration.

Perhaps the most widely accepted contemporary definition or statement on greenways in the USA was included in the report of the President's Commission on Americans Outdoors in the USA (1987). The Commission advocated a greenways network:

> to provide people with access to open spaces close to where they live,
> and to link together the rural and urban spaces in the American

landscape threading through cities and countrysides like a giant circulation system.

This statement emphasises the concept of spatial connectivity, of an integrated functional network, managed for multiple purposes, linking rural and urban environments. It also reflects the late twentieth century orientation in contemporary American land preservation, which focuses on open lands that are directly accessible to population centres, in contrast with the nineteenth and early twentieth century emphasis on the great, but more remote, national parks and other protected landscapes (Zube 1996).

A comprehensive set of definitions on greenways was provided by Charles Little, the author of the popular 1990 book *Greenways for America*. In his book a 'Greenway' is defined as:

1. A linear open space established along either a natural corridor, such as a riverfront, stream valley, or ridgeline, or overland along a railroad right-of-way converted to recreational use, a canal, scenic road, or other route.
2. Any natural or landscaped course for pedestrian or bicycle passage.
3. An open-space connector linking parks, nature reserves, cultural features, or historic sites with each other and with populated areas.
4. Locally, certain strip or linear parks designated as parkway or greenbelt.

(Little 1990)

Little's definition shares the same fundamental ideas as the President's Commission; in addition, it recognises specific types of greenway depending on their location, spatial configuration and purpose. Many other authors of greenway books, journal articles and reports cite the definitions of Little and the President's Commission (Smith and Hellmund 1993; Flink and Searns 1993; Erickson and Louisse 1997).

I proposed another greenways definition in the book *Greenways: the Beginning of an International Movement*, based on literature review and research and applications experience with Greenway planning projects in the USA:

Greenways are networks of land that are planned, designed and managed for multiple purposes including ecological, recreational, cultural, aesthetic, or other purposes compatible with the concept of sustainable land use.

(Ahern 1996)

This definition is intended to be comprehensive and inclusive. In this cited reference, I also propose a typology of greenways, structured to enable explicit,

Table 3.1. *Linkage of greenway definitions with theoretical principles*

Definitional themes	Questions/issues raised?	Theoretical principles
Greenways are a linked, or spatially integrated network of lands that are owned or managed for public uses including: biodiversity, scenic quality, recreation, and agriculture	Does this network produce an advantage due to an intrinsic pattern of resource distribution?	1. Hypothesis of co-occurrence of greenway resources
A presumed advantage, or synergy, resulting from spatial connectivity and linkage	How do the determining functions affect the spatial form and configuration of the greenway?	2. Inherent benefits of connectivity for humans and for biodiversity
	Is there sufficient knowledge and information available to plan for connectivity?	
Planned to accommodate multiple uses and to achieve multiple goals	If the greenway is multi-purpose and multi-objective, which are the primary or determining uses/functions?	3. Compatibility and synergy of multiple use(s)?
	Are the uses spatially compatible or conflicting?	
	Who decides which uses take priority? Is the greenway spatial configuration intentional and deliberate, or is it opportunistic?	

comparative description and communication of greenways across physical, spatial, cultural and political contexts. The typology classifies greenways according to: spatial scale, purpose and goals, landscape context, and planning strategy (Ahern 1996).

These definitions illustrate the diversity that is inherent in greenways in concept and in reality. Perhaps this diversity helps to explain the popularity of greenways, and also emphasises the need for a greenway classification or typology to assure a clear and accurate communication between researchers and professionals. Table 3.1 relates greenway definitions with significant questions raised. The definitions and questions lead to three theoretical principles that are posed as representing a theoretical basis in support of greenways: (1) the hypothesis of co-occurrence of greenway resources; (2) the inherent benefits of connectivity; and (3) the compatibility and synergy of multiple use in greenways. The following section (3.2) discusses and explores these three greenway principles, which represent an emerging theoretical basis in support of greenways.

3.2   Greenway theory

3.2.1 Hypothesis of co-occurrence of greenway resources

One of the common arguments in support of greenways is the hypothesis of co-occurrence of greenway resources. When discussing greenway resources, it is important to distinguish from earlier, conventional conceptions of protected landscapes. The USA's National Park system is well known for its spectacular natural scenery, typically remote from urban regions. Greenways embrace the concept of protected lands within urban regions, explicitly and intentionally located in close proximity to where people live and work. Greenway resources thus include the riparian/drainage network, large patches, small 'bits of nature', and linking corridors (Forman 1995). The hypothesis of co-occurrence posits that in any cultural landscape greenway resources are spatially concentrated along corridors. Cultural landscapes in the USA are understood differently from those in Europe. As a younger culture, American concepts of cultural landscapes are still emerging, rooted in the traditions of colonial agriculture, vernacular rural and suburban landscapes, and greenway corridors typically include riparian and linear upland areas, such as regional topographic ridges and small mountain ranges. Although the hypothesis warrants further and continued testing, several investigators' results support this hypothesis across a range of scales and contexts in the USA and in Europe. If the hypothesis is valid, greenways offer three strategic advantages:

- spatial efficiency: because they consist largely of corridors, where resources are concentrated, greenways can protect the most resources with the least amount of land area
- political support: political consensus and support is more likely to occur owing to the mutual benefits that diverse interests can realise from greenway protection (e.g. recreational, biodiversity, water quality)
- connectivity: if greenway resources are concentrated in corridors, the benefits of connectivity will be expressed in ecological, physical and cultural terms (see section 3.2.2 for a discussion of the inherent benefits of connectivity)

The earliest research on the hypothesis of co-occurrence of greenway resources is usually attributed to Philip Lewis, a landscape architecture professor and practitioner from Wisconsin, USA. Lewis' classic study for the Wisconsin Outdoor Recreation Plan surveyed and mapped the locations of 220 ecological, recreational, cultural and historic resources. Lewis' study found that over 90% of these resources occurred along corridors, which he labelled 'environmental corridors' (Lewis 1964). These corridors were used as the basis for the Wisconsin Heritage Trail Proposal (Figure 3.1). Lewis' work is well known in the USA as a precursor to modern greenways. Lewis recognised the importance of this co-occurrence, not only as a means towards efficient land protection, but also to show diverse public constituencies that their respective interests are often spatially coincident. The environmental corridors have also proven important for education by increasing awareness of connections among a variety of natural and cultural resources that tend to co-locate along greenway corridors (Lewis 1996).

A more recent study in the state of Georgia, USA, produced findings that also support the hypothesis of co-occurrence. The 1976 Environmental Corridor Study by the Georgia Department of Natural Resources included an extensive, statewide inventory of intrinsic (natural) and extrinsic (social) landscape resources (Dawson 1996). The study's research method included four steps:

- resource analysis
- corridor selection and priorities
- corridor planning and management options
- summary and conclusions

The resource analysis was followed by a series of assessments and map overlays that identified the preliminary corridors where the most significant greenway resources were located. These corridors became the priorities for greenway land acquisition. The mapped concentrations of greenway resources led to a statewide greenway plan, which has since begun to be implemented.

Since 1994, the Metropolitan Region of Lisbon, Portugal, has been developing a greenway plan (Machado *et al.* 1995). The plan has developed according to a broad and inclusive understanding of greenway resources, both natural and cultural. This work builds on the earlier work of Gonçalo Ribeiro Telles, in his 'Continuum Natural' and the more recent 'Plano Verde de Lisboa', which articulated a continuum in which the spatial distribution of natural and cultural resources can be understood in a cultural landscape (Telles 1975, 1997). Telles' work anticipated the greenway concept, and identified the importance of green corridors, where resources are concentrated, to link natural and cultural landscapes in the region, including the city of Lisbon.

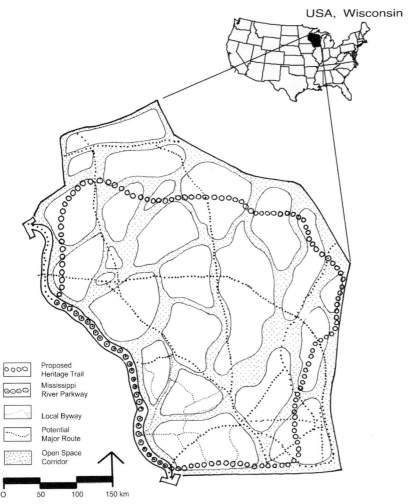

FIGURE 3.1.
Lewis' Wisconsin Heritage Trail Plan includes over 300 km of Environmental
Corridors where ecological, recreational, cultural and historic resources are
concentrated. (From Fabos 1985, p. 118.)

In the first phase of greenway planning for the Lisbon Metropolitan Area
(Área Metropolitana de Lisboa, AML), broad scale spatial databases were used to
identify 'greenway corridors' where natural resources were expected to be concentrated. The corridors were defined as coastlines, river and stream valleys,
and major ridgelines. Several nationally significant and one UNESCO World
Heritage Landscape (Sintra) are located in the AML region. Through spatial
overlay analysis these cultural resources were found to co-occur within the
'greenway corridors' defined. A large gap in the data existed, however, because
the spatial locations of cultural resources were not available for GIS (geographical information systems) analysis. Working with the Portuguese National GIS

agency (CNIG), Ribeiro (1998) compiled a spatially explicit database of over 3000 cultural resource sites in the North Bank of the Lisbon Metropolitan Area (AML). Ribeiro's analysis not only identified the corridors in which the significant cultural resources were located, but also articulated the causal linkage with the natural features and regions that influenced the historical development of these cultural resources, and pointed towards strategies for their interpretation in a regional greenway plan. His work also verified the hypothesis of co-occurrence of greenway resources at the regional scale, with particular emphasis on the spatial distribution of cultural greenway resources.

The Minute Man National Historic Park in Massachusetts, USA, provides another examination of the hypothesis of co-occurrence of resources in greenways. This park was established along a linear corridor that was determined and delineated for its historical and cultural significance relating to an early battle in the American Revolutionary War. In the context of a multi-purpose planning exercise, it was learned that a very significant concentration of biological (rare species habitats, unique or rare ecosystems) and abiotic greenway resources (prime agricultural soils, mature vegetation patches) occurred along the corridor. Subsequently a greenway plan was developed to delineate an interpretive route with the goal of interpreting both the natural and cultural history of the landscape (Gavrin et al. 1993; Ahern 1996).

These selected cases illustrate where the co-occurrence of greenway resources has been demonstrated through planning projects of differing scales and contexts. No contradictory findings were identified in a thorough review of greenway literature. However, it is recognised that additional research, structured by a clear hypothesis and executed in accordance with a consistent and replicable method, would yield a necessary examination of the hypothesis. For the purpose of this chapter, it is assumed that the hypothesis of co-occurrence is a reasonable and valid working hypothesis.

3.2.2   Inherent benefits of connectivity

Connectivity is defined here as a spatial characteristic of systems (i.e. landscapes) which enables and supports the occurrence of specific processes and functions, through adjacency, proximity or functional linkage and connection. The sustainability of certain landscape processes is dependent on connectivity. These processes include, for example, the movement of wildlife species and populations, the flow of water, the flux of nutrients, and human movement. Given this definition, it is argued here that providing or maintaining connectivity in a landscape supports particular processes and functions that may not otherwise occur. If these processes are beneficial and valued by humans, and are dependent on connectivity to some extent, then it can be argued that connectivity is an important characteristic of, or a prerequisite for, sustainability.

The nature of the 'connection' implicit in the term connectivity is a function of the process or function that is being supported. For the flow of water, for example, a continuous, physically linked system is needed, because water moves according to physical laws under the influences of gravity and topography. For wildlife movement, the nature of connectivity is species-dependent. As conscious, mobile organisms, wildlife species demonstrate preference for, or avoidance of, certain landscapes or landscape features (Bennett 1998; Forman 1995). Some species (e.g. birds) can move across great distances between habitat patches using intermediate 'stepping stones', whereas other species (e.g. mammals) are often dependent on a physical corridor connection to facilitate movement. A habitat network for birds, then, may consist of a series of patches, each separated by kilometers of unconnected landscape. However, when the distance between the 'stepping stones' becomes too great, connectivity ceases to exist. A corresponding network for aquatic mammals (e.g. the river otter, *Lutra canadensis*), needs to have a virtually continuous, physically linked habitat. Connectivity must be understood in terms of the process or function that it is intended to support (Bennett 1998; van Langevelde 1999).

A great deal of the literature of landscape ecology addresses the inherent value of connectivity with respect to biodiversity (van Langevelde 1999; Bennett 1998; Forman 1995; Vos and Opdam 1993; Saunders and Hobbs 1991; Soulé 1991b; Turner 1989; Schreiber 1988). Much of this literature focuses on the importance of connectivity for maintaining biodiversity in landscapes that are urbanising, or otherwise experiencing a reduction in area or a fragmentation of species habitat. This argument has been criticised by others who maintain that the benefits of connectivity have yet to be scientifically established, that connectivity may, in fact, inadvertently enable the spread of disturbance, disease and invasive species, and that conservation funds may be more wisely spent on the acquisition of habitat patches (Simberloff and Cox 1987; Hess 1994). In a review article, Beier and Noss (1998) articulate a position, based on review of the empirical research, which supports the value of connectivity for habitat corridors in biodiversity protection as follows:

> The evidence from well-designed studies suggests that corridors are valuable conservation tools. Those who would destroy the last remnants of natural connectivity should bear the burden of proving that corridor destruction will not harm target populations.
>
> (*Beier and Noss 1998*)

This argument is consistent with the 'Precautionary Principle' contained in the Rio Declaration on Environment and Development, the comprehensive international policy statement that supports the international goal of sustainability:

> In order to protect the environment, the precautionary approach shall be widely applied by States according to their capabilities. Where there are threats of serious or irreversible damage, lack of full scientific certainty shall not be used as a reason for postponing cost-effective measures to prevent environmental degradation.
>
> *(IUCN 1992, Principle 15)*

The value of ecological corridors has been widely accepted in recent European conservation planning and policy. The European Union's Habitats and Species Directive, adopted in 1992, proposes connectivity via corridors and stepping stones to link and to assure favourable status for special areas for conservation (SACs). The directive includes the plan 'NATURA 2000', which identifies the core areas and linkages necessary to maintain favourable status for the SACs:

> European Union Member States should endeavour in their land-use planning and development policies to encourage and manage features of the wider landscape which are of importance for wild fauna and flora. Linear features, such as rivers and hedgerows, and isolated elements, such as lakes and ponds, are essential for migration, dispersal and genetic exchange of wild species.
>
> *(EU Habitats and Species Directive art. 10, 92/43/EEC)*

The more recent Pan-European Biological and Landscape Diversity Strategy of 1995 (Council of Europe 1996) was prepared to enable implementation in Europe as it has been signed by 54 countries and is supported by the European Union. The strategy specifically proposes ecological network elements such as corridors, buffer zones and stepping stones to reduce the effects of isolation and to increase viability for small areas (Nowicki *et al.* 1996). This strategy explicitly addresses cultural and economic issues as part of the planning context. In this manner, they are similar to greenways in spatial organisation, and function. The spatial planning response to these policy directives and strategies are most often labelled ecological networks, often with rivers and streams as their principal spatial organising elements:

> An ecological network is successful if it sustains biological transition and landscape connectivity at all levels where fragmentation, isolation, and barriers to movements and fluxes occur. Rivers and water flows in

general can play an important role in this because of their function in supplying water and transporting sediments, nutrients, and organisms.

(*Jongman 1998b*)

Bennett (1998) points to a common confusion between connectivity and corridors, noting that connectivity can be achieved in some landscapes without 'corridors' *per se*. Others define this distinction as functional versus actual connectivity (van Langevelde 1994). This broader concept of connectivity in landscapes is a characteristic that is more widely accepted, and that is compatible with the greenway concept. National and international plans have been made, and are being implemented based on the importance of connectivity (Nowicki *et al*. 1996). The precautionary principle supports this approach.

The time scale needed to scientifically test the efficacy of habitat corridors in large landscapes is decades or centuries. In the time that would elapse during such a study, most landscapes would have changed fundamentally in terms of structure and function. This 'moving target' for research creates a fundamental dilemma for landscape planning. How can plans be made to address contemporary concerns and short-term goals with incomplete or imperfect knowledge? The concept of adaptive planning or management offers a conceptual solution to this dilemma. It is a flexible scientific framework for re-conceiving landscape plans, or management actions, as experiments which may, over time, yield new knowledge regarding the effectiveness of the plan or action. The adaptive approach is well suited to testing the efficacy of corridors in varying landscape contexts and for differing purposes (Peck 1998). A rigorous application of the adaptive planning approach is dependent on a proper monitoring protocol, adequate data, a robust analytical design, and a mechanism for the incremental knowledge gained to influence the future planning and management of the landscape in question.

Connectivity in hydrological systems is a key attribute. Water flows across landscapes under the influence of gravity, influenced by vegetation, geology, topography, and human engineering. The physics of hydrological flows across landscapes over time results in the formation of discrete channels and stream networks. These networks have been described as the river continuum, in which hydrological, physical and biological processes and structure change according to the position in the watershed (catchment) (Vannote *et al*. 1980; Naiman *et al*. 1987). For example, in the headwaters (i.e. low-order streams) food chains are based primarily on detritus, water temperature is cool, and stream flow rapid. In the lower sections of the watershed (i.e. higher-order streams), the food chain is based on micro- and macro-invertebrates, water temperature is warmer, and flows slower. The channels of the river continuum, and their associated border zones of hydrological influence, are together defined as riparian corridors, which in most landscapes support a distinct floodplain vegetation adapted to the soil, hydrology and disturbances characteristic of the riparian zones. Riparian corridors contain important longitudinal and transverse ecotones (Pinay *et al*. 1990). The functions of these riparian ecotones include movement and retention of nutrients, exchange of organic material, and development of floodplain vegetation. The physical and functional connectivity inherent in riparian systems supports movements of materials and organisms between the main fluvial channel and its secondary channels and oxbow lakes. Through this process, nutrient spiralling, downstream nutrients are absorbed, utilised and released by organisms. Nutrient spirals are linked to regulate longitudinal movement and retention of nutrients in fluvial systems. These movements are critical for fish habitat. These functions are susceptible to interruption through dam or dike construction, which can reduce or eliminate the riparian zone bordering the channel and may interrupt the movement of materials and nutrients downstream.

The riparian zone contains, by definition, the zone of intersection of the surface and subsurface hydrological systems. Riparian corridors are fundamental to greenways for they provide connectivity, contain many resources, and support multiple uses and functions. Forman (1995) argues that riparian corridors anywhere in the world are 'indispensable' for the sustainable functioning of any landscape because the functions they provide cannot be provided by any other means or location in a landscape.

The laws of physics dictate that hydrological systems cannot be interrupted, for the water that flows downstream must be conveyed in a channel, or it will form a new channel. Human disturbance in riparian zones tends to constrict the channel, and 'disconnect' the riparian zone from the channel. In this common case, the channel provides only a conveyance function. The focus of much greenway work is to restore riparian zones along channels, thereby supporting the other functions and processes that occur when a continuous riparian zone exists. These collateral functions include stabilising surface and ground-water flows (recharge and discharge), wildlife habitat and movement corridors, nutrient and sediment buffering, human recreation, and support for cultural landscapes. The width of riparian corridor required to support these functions will vary as a function of the order of the stream channel (i.e. watershed position), the degree of human hydrological control, and hydrologic flow and disturbance regime (Forman 1995).

Other benefits of connectivity that can be supported by greenways include alternative forms of transportation, trail recreation, and the human need or preference for nearby nature and recreation (Kaplan *et al*. 1998). Comprehensive state-wide greenway plans have been developed that integrate these benefits explicitly (Florida Greenways Commission 1994). Transportation, by definition, involves locomotion between an origin and a destination.

Connectivity is essential for transportation to function. Greenways are often planned and implemented to support alternative forms of transportation, particularly pedestrian and bicycle travel. The benefits of this may be significant in terms of traffic reduction, reduced air pollutants, and a healthier population. In many regions of the USA, continuous, integrated bicycle trails are unusual, but when provided, prove to be immensely popular (Flink and Searns 1993). The same kinds of benefit occur from trail linkages. When greenways provide walking trail connections with other trail systems, their level of use and value is increased. Many greenways in the USA begin as single-purpose trail systems.

Perhaps the most abstract benefit of greenway connectivity is the psychological one – of linking people with nature, close to where they live and work. This goal is reflected in the President's Commission Report (1987). Kaplan *et al.* (1998) address the human need and preference for 'nearby nature', to experience the natural world as matter of course in everyday life. Providing this benefit has been shown to improve personal and social health. Historically, human–nature interaction was provided through large parks in cities (Fabos 1995). The greenway concept brings a new strategy to bear on this issue. By establishing 'fingers of green' in the urban and suburban areas where people live, a physical connection is made which supports this philosophical or spiritual need for human–nature contact. When the human–nature access links with other resources, the benefits are multiplied in a synergistic manner, at least in terms of space utilisation. This provision of multiple benefits is the subject of the next section.

3.2.3 Compatibility of multiple use

The final assertion of this section on supporting theory states that greenways are viable because they provide multiple functions within a specific and often limited spatial area, and that these uses can be planned, designed and managed to exist compatibly or synergistically. This argument is presented in three parts: (1) presumption of compatibility, (2) economic benefits, (3) building a base of political support through multiple use.

Presumption of compatibility

The claim that multiple uses can exist within a corridor of protected land presumes some degree of compatibility between the uses, for if the combination of two or more uses compromises the value or function of all, then no net benefit is gained. Testing this presumption is possible when the spatial requirements of the combined functions are well known. For example, protected wooded riparian corridors can provide emergency flood control function and routine recreational and scenic uses. In this case there is an inherent compatibility between the uses, with only an occasional disruption during periodic floods. A more complex, but common, combination occurs when wildlife habitat functions are integrated with recreational access in greenways. Few species' spatial and habitat requirements are sufficiently well understood to be represented and modelled in a spatially explicit manner in greenway planning. Most species are not understood to this degree, making habitat planning a complex and uncertain process. Further, it is important to acknowledge that all habitats are not equally adaptable to multiple use. For example, forest-interior, disturbance-sensitive species are difficult to integrate into a greenway plan. When greenway management permits, timing of recreational access can reduce the impact of human presence and disturbance by managing the time, place, and intensity of the use. Examples include restricting access during nesting or breeding periods, or restricting access to forest patch margins or perimeter areas.

Economic benefits

The spatial efficiency inherent in the co-occurrence of greenway resources has an economic dimension. When multiple functions are provided in a single corridor, less land is pre-empted from other uses for these purposes. In addition, there are economic efficiencies in land acquisition, planning, design and management costs and expenses.

The broad economic benefits of protecting land for public use (including greenways) have recently been summarised by the US Trust for Public Land (TPL) (Lerner and Poole 1999). The TPL's report, 'The Economic Benefits of Parks and Open Space' identifies three categories of economic benefits related to land protection. Firstly, there is the 'Smart Growth' argument. Open land protection promotes more concentrated development patterns, thereby reducing the costs of providing infrastructure for low-density or sprawl-type development. This is a timely issue in the USA, where unplanned, decentralised urban development, or 'sprawl', is an important issue on the national agenda. Planning for integrated, linked protected lands within new urban developments is promoted as a prime 'quality-of-life' issue. Numerous studies cite access to natural areas and recreation as primary factors in people's preference for residences (Lerner and Poole 1999).

Secondly, open land protection promotes many forms of economic activity and investment. Parks and open space attract business and residents to communities, stimulating commercial growth, tax revenues and tourism. In many regions tourism is fast becoming the main economic activity. Open space is now recognised as an integral component of a sustainable economy.

Finally, open land protection provides a cost-effective means to safeguard the environment, producing a direct benefit for humans. These beneficial functions include: flood protection, water storage and purification, air cleaning, degradation of organic wastes, and reducing urban heat island effects. These economic benefits can be attributed to any form of protected land, including greenways.

Building a base of political support through multiple uses
When greenways are integral to urban development, the opportunities and challenges for compatible multiple use come to the forefront (see Box 3.1). Clearly, choices and trade-offs need to be made to optimise any particular use. As these trade-offs become more explicit and intentional, new knowledge can be generated through monitoring and continued evaluation and research.

BOX 3.1. Guidelines for providing multiple functions in greenways

The test of compatibility of multiple uses is fundamental in greenway trails. New expertise is developing regarding the design of greenway trail corridors to support multiple functions, particularly wildlife habitat. The following is a summary of the key emerging concepts relating to greenway trail planning, design and management.

A. Understand trail impacts. Greenway trails have specific zones of influence, which need to be planned with awareness of the timing, nature and intensity of trail use, and with the nature of the landscape through which the trail passes.
B. Plan greenway trail routes carefully. Greenway trails should avoid crossing large natural areas. They should follow, not create, disturbance zones around protected core areas. Overall trail density should be kept as low as possible.
C. Understand trail users. Trail users are a diverse and heterogeneous group, each with unique and important characteristics. For example, humans, dogs, and horses may all use a trail, and each has particular needs and impacts.
D. Manage trail use. Greenway trail management is an ongoing process, which needs to employ a full range of management actions including trail closure, limits of use, and trail repair and restoration.
E. Monitor trail impacts over time. Begin with an initial biological inventory, followed by monitoring. Enforce trail closures.
F. Involve users and the public with trail management. Develop a sense of stewardship and engage volunteers in trail planning, implementation and management.
(Adapted from Hellmund Associates 1998; Smith and Hellmund 1993; Flink and Searns 1993.)

The greenway movement has been criticised for following a parochial 'parks and recreation' focus. Although this historical orientation produced many notable results and successes in rural areas, it reached its limits in urban areas. When greenways are conceived to provide multiple benefits, they hold the potential to engage multiple political constituencies in their implementation. This has proven to be an effective strategy in successful implementation of greenways in multiple cases across the USA (Erickson and Louisse 1997; Quayle 1996). In the USA the tradition of planning for land protection is much more developed in remote and isolated, spectacular landscapes than in urban centres and regions, particularly within metropolitan areas.

Once realised and implemented, multi-purpose solutions hold a greater potential to endure over time, as demographics, economics, environmental issues and landscape context change. In this respect, it is interesting to compare multi-purpose greenways with greenbelts. The former are inherently multi-purpose, the latter tend towards a single purpose, i.e. to contain urban expansion around urban areas. The greenbelt concept, because it was based on political boundaries and not natural features, became vulnerable to land use change, effectively becoming a 'bank' in which undeveloped land was held until development pressure demanded its use. In Canada's capital city of Ottawa, a greenbelt was established in 1950 as part of a regional plan. The greenbelt was incrementally compromised as pressure for land use change mounted over time (Taylor *et al.* 1995). In contrast, greenways that support multiple functions inherently enjoy a broader base of political support, and therefore can be more sustainable over time.

3.3   Historical development of greenways in the USA

A brief historical review of greenways in the USA illustrates an evolutionary process which parallels and reflects innovations in American landscape planning. In the public domain, vast areas of public land were added to the US National Park and US Forest systems over the past two centuries. These actions involved large pristine areas, far from human populations, which came under public control. Despite the monumental amounts of land involved, this was a relatively easy task because much of the land was never released from federal control, and much of it is mountainous and arid, thus less suitable for agriculture or urban uses. Greenways are quite different from these national parks for they focus on linear areas, are more often located near population centres, and are managed for multiple uses. This review articulates the issues, theories and policies that led to the development of greenways and serves as a basis for understanding their present situation and potential for future development.

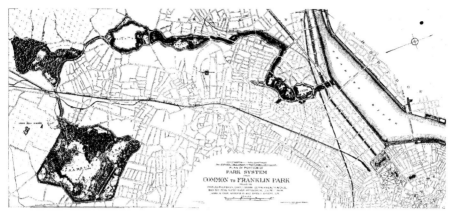

FIGURE 3.2.
Frederick Law Olmsted's 'Emerald Necklace' plan for the Boston Park System, the
first greenway in the USA, 1880s. The black structure represents the Emerald
Necklace, totally situated within the urban area of Boston.

Most of the literature on greenways points to their evolution from urban design concepts of the nineteenth century, including boulevards, axes and parkways. Searns (1996) labels these as first-generation or ancestral greenways. The
first true greenways originated from the metropolitan open space systems of
the late nineteenth and early twentieth centuries (Fabos 1995; Newton 1971;
Zube 1996; Smith and Hellmund 1993). These were 'systems' in the sense that
they involved a spatially linked network of mostly linear publicly owned lands.
They were usually based on topographic and hydrological patterns in the landscape. Foremost among these systems was a built plan for the Boston Park System by Frederick Law Olmsted, Sr. (c. 1880s), the father of landscape architecture in America. His later work involved his sons as partners (Zaitzevsky 1982).

The Olmsteds' plan for the Boston Park System, known as the 'Emerald
Necklace', is regarded as a model of integration of existing protected lands,
ecological corridors, and built linear elements (Figure 3.2). The system largely
functions today to provide recreation, transportation, water quality and flood
control, scenic amenity, and wildlife habitat. The 'necklace' is a fine example of a simple, yet powerful and enduring spatial planning concept. Under
Olmsted's hand, several other American cities embraced this concept of linked
linear parks, including Washington, D.C., Minneapolis, Kansas City, Buffalo
and Cleveland.

The next historically significant greenway was the Metropolitan Boston
Park System in the 1890s, planned by Charles Eliot, a protégé of Olmsted
(Figure 3.3). Eliot's work greatly expanded Olmsted's 'emerald necklace', by
creating a regional open space system, or greenway, structured by five principal landscape types which closely resemble contemporary greenway elements:

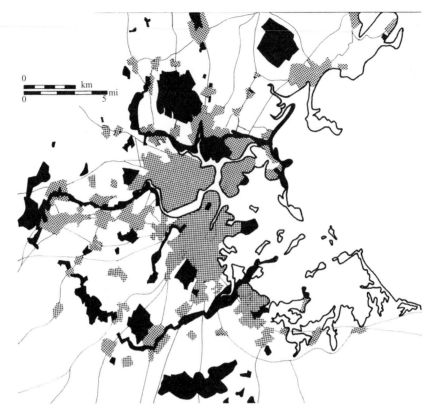

FIGURE 3.3.
Charles Eliot's Metropolitan Boston Park System, *c.* 1899. (Courtesy of Fabos 1985.)
Black, Metropolitan Park System; grey, built-up areas.

ocean fronts, river estuaries, harbour islands, large forests, and small urban squares.

Benton MacKaye (1928) expanded Olmsted and Eliot's urban park system concept in his book *The New Exploration*, in which he advanced, for the first time, the idea of a metropolitan system of protected lands conceived and configured to control urban expansion. Influenced by the earlier work of Ebenezer Howard in England, using the analogy of a river, he identified topographic ridges as 'levees' to contain and control the 'flow' of metropolitan urban expansion. The environmental planning movement of the 1960s marks the next significant development towards greenways. Ian McHarg's *Design with Nature* (1969) raised international awareness of the need for an ecological basis for planning and advanced a widely adopted method to accomplish it. McHarg argued that the major landscape planning issue was that of influencing the pattern of distribution of occupied and protected lands, not their absolute or relative areas. The work of Phil Lewis, mentioned previously, integrated environmental planning through his 'Wisconsin Heritage Trail Proposal' (Figure 3.1).

The greenline concept of the 1970s introduced a new idea in land protection and management based on mixed public–private ownership. Greenline parks are mixed mosaics of public and private lands not defined exclusively by public ownership but rather by a 'green line' on a map. This idea, based on the national parks of England and Wales, responded to decreased federal funding for land acquisition, and the awareness of the need to protect open space within urban and metropolitan areas (Zube 1995). The greenline parks were represented by the Adirondack Mountains in New York, the New Jersey Pinelands, and many urban recreational areas within or adjacent to major cities.

As the concept of greenline reserves evolved, its emphasis shifted from large, park-like reserves to linear corridors including historic canals, railroads, and rivers. This was largely due to an emphasis on riparian corridors responding to the unprecedented expenditure made by the federal government to clean America's rivers in the 1960s. The effort brought attention to the problems of water pollution, and then when the rivers were once again clean, their recreational potential was rapidly rediscovered and developed. The Wild and Scenic Rivers Act of 1968 provided additional protection for rivers, wetlands and coastal zones, adding further interest to innovative models for the protection of linear landscape features. There was an emphasis on trail-oriented recreation during this period, which Searns (1996) labels the second generation of greenway evolution. Greenline reserves have evolved further into a planning–management entity known as National Historic Corridors (NHC), of which there are currently over 15 in the USA. NHCs are essentially greenways because they consist of linear areas, are spatially integrated and are managed for multiple uses. These third-generation greenways (Searns 1996) are truly multi-objective and demand an interdisciplinary planning and design approach.

In the 1980s, the loss of open space and increased need for recreation in urban and metropolitan areas focused attention on greenways. The President's Commission on Americans Outdoors (1987) found strong support for greenways to address the need for additional open space and recreational land and proposed a national system of greenways (see introduction).

Greenways have evolved into a flexible multi-purpose model for landscape planning and resource protection. In contrast with the 'crown jewels' of the US National Park System (e.g. the Grand Canyon, Yosemite), greenways protect 'working' landscapes in cities and regions where people live and work. The strategic nature of greenways suits them well to situations where land use must be spatially efficient and multiple uses are essential to gain political and economic support. The continuing evolution and adaptation of greenways is discussed in the following section.

3.4   Contemporary trends

Although US greenway history can be traced back over one hundred years, it is clear that greenway activity has never been as effective as in the contemporary time. An examination of this recent greenway activity is useful to understand the continuing evolution of greenways, to identify opportunities for applications and future research needs.

To gain a more accurate understanding of the nature, extent and location of greenway planning across the USA, a national survey of greenways was conducted by the University of Massachusetts, from 1996 to 1998. The survey was designed with the primary goal of identifying the nature and extent of greenways and greenway planning, at the state level, across the USA. The survey was conducted with assistance from the American Greenways Program, sponsored by the Conservation Fund (Washington, D.C.). Officials from each state were given a standardised telephone interview followed with a written request for information and responses to specific questions.

The survey found that whereas 48% of the states supported the concept of greenways, an equal number (48%) were not familiar with greenways at all. Not surprisingly, therefore, only 24% of the states had an official greenway plan, with 68% indicating that there was no plan. A similar response was obtained regarding the existence of a mapped inventory of greenways: 24% responded yes, 62% no and 14% did not know whether such an inventory existed for their territory. The survey also found a trend regarding the spatial distribution of greenway planning across the USA. Greenways were found to be most popular in the east and northeast where the states are small, population density is high, and the percentage of publicly owned land is low. The results of this survey also indicated that greenways are often initiated to support trail and recreational use, but evolve to support multi-purpose planning objectives. Finally, the survey found that greenways are increasingly integrated with comprehensive statewide planning.

The Rails-to-Trails Conservancy (RTC) has been active in greenway and trail creation since the 1980s. In the decade from 1988 to 1998, RTC helped to convert over 10 000 miles of abandoned railroads to greenway trails. Another recent greenways-related activity occurred in 1999 when the White House, the US Department of Transportation and the Rails-to-Trails Conservancy established the Millennium Trails Program. The goal of the programme is to recognise, promote and stimulate the trail movement in the USA and to reconnect communities with trails. Under this initiative, trails are designated in three categories: National Millennium Trails (12), Millennium Trails (52) and community trails (thousands). This programme will advance trail-based greenways, increase public awareness of all greenways, and most likely inspire future greenway development.

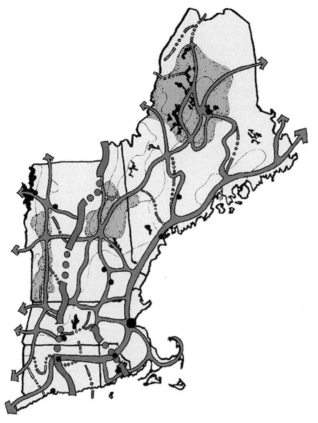

FIGURE 3.4.
New England Greenway Vision Plan (Fabos *et al.* 1999), showing the main greenway network.

At the regional level, the most significant greenway planning effort to date is the New England Greenway Vision Plan (Fabos *et al.* 1999). The plan co-ordinates greenway planning for all six New England states with a combined land area of over 42 million acres (*c.* 16 million hectares) (Figure 3.4). The plan, prepared in collaboration with the American Society of Landscape Architects, builds on the tradition of Frederick L. Olmsted Sr, Charles Eliot, and Benton MacKaye, who worked extensively in New England. The Plan was prepared through a co-ordinated, decentralised effort which integrated locally with statewide plans. Using GIS, the vision plan integrates single-purpose plans for nature protection, recreation and historic and cultural resources. The plan emphasises linear features, the importance of connectivity, and the imperative for multiple uses.

3.5   Conclusions and prognosis

Greenways represent an efficient and strategic method for protecting the most resources with the least amount of land (the hypothesis of co-occurrence).

The connectivity inherent in greenways supports numerous biological, physical, and cultural landscape functions that are important for sustainability. Within a spatial network of important and connected lands, greenways are planned and managed to support multiple compatible uses, thus assuring a broader base of political support and increasing the probability that the greenway lands may remain under protection for the future.

Greenways are becoming a popular international movement (Fabos and Ahern 1995). The theory of greenways and greenway planning presented in this chapter provides a rational basis for supporting greenways, and identifies future opportunities for application, and need for additional research. It is argued that these three ideas provide a rational basis of support, supported by published literature and case studies that transcend political context or geographic location.

Greenways are a strategic planning concept that has evolved over the past century in the USA in response to changing environmental, cultural, political and economic factors. The proliferation of decentralised, urban sprawl has motivated interest in alternative planning models and methods. The greenways concept addresses this need through its strategic approach, its record of successful integration of top-down and bottom-up approaches, and its emphasis on physical and organisational linkages. Physical linkages offer distinct advantages in terms of movement and transport of materials, species or nutrients. Greenways also provide a social and political network, which integrates people with diverse values and perspectives concerning land use and planning. This is perhaps the most significant characteristic of greenways, and distinguishes greenways from other landscape planning concepts.

The continued evolution and implementation of greenways is likely to produce three significant changes in the future.

Greenway planning will change the ways in which local and higher-level planning is co-ordinated and implemented. A major driver of this change is the emphasis on corridors, and the assumed value of connectivity, which together emphasise the need to link local plans to their larger landscape and regional context. This co-ordination will be accomplished, in part, through a new suite of land use controls and design guidelines in lieu of public ownership of land.

Greenways will inspire and motivate a new generation of partnerships and collaborations among individuals and organisations that formerly had some common interests, but with little record of co-operation (e.g. wildlife habitat and recreation, tourism, water resources). As Zube (1996) stated, 'Partnerships are a way of life in greenways.' Formal agreements for planning and technical assistance and for interagency and intergovernmental co-ordination will become more the norm than the exception.

Finally, greenways will promote an adaptive approach to the dilemma of landscape planning and management. Although greenways will continue to apply the best available empirical knowledge and theory from landscape ecology in decision making, this knowledge, with respect to specific places and processes is inherently uncertain and incomplete. At the same time, social, political and environmental changes demand that actions be taken. The adaptive approach to planning and management offers a solution to this dilemma. Planning and management decision can be re-conceived as experiments, with the potential to add new knowledge as a result of their application. Greenways are well suited to this adaptive approach.

4

Ecological functioning of ecological networks: a species perspective

4.1   Introduction

One of the main functions of ecological networks is to protect and enhance biodiversity. If we want to know whether conservation and restoration plans meet our biodiversity goals, we should be able to assess them. Over the years, research on the functioning of ecological networks and investments in assessment tools have made it possible to do this. In this chapter we combine the Landscape Cohesion (LC) approach developed by Opdam *et al.* (2003) with Ecologically Scaled Landscape Indices (ESLIs) (Vos *et al.* 2001) and spatial sustainability standards (Verboom *et al.* 2001). The approach is based on the functions of an ecological network for the organisms that inhabit it. In this approach, ESLIs are compared to standards resulting in a rating of the landscape. This rating can be viewed as an index of landscape ecological spatial quality (Figure 4.1). A complication when assessing biodiversity is that every species perceives a landscape from its own perspective. Therefore, the basis of the LC approach is to determine species-specific ratings, and then combine them into overall, general ratings.

The most important ecological concept for assessments in human-dominated landscapes in general, and ecological networks in particular, is the metapopulation concept (Levins 1970; Opdam 1988; Hanski and Gilpin 1997). A metapopulation is a set of populations in a habitat network connected by inter-patch dispersal. A habitat network is a set of habitat patches close enough to have a reasonable level of inter-patch dispersal. Habitat is a species-specific term for the set of conditions a species needs to feed, survive and reproduce. In this chapter we present a method of looking at ecological networks in a functional way, using our knowledge of the population processes in habitat networks, i.e. local extinction and patch colonisation, for prediction of sustainability of the landscape. The key of the approach is to perceive the

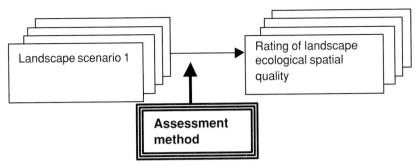

FIGURE 4.1.
Visualisation of the problem of labelling a landscape plan or scenario by rating (quantitative index), assessing the spatial quality based upon knowledge of landscape ecological processes. The assessment method discussed in this chapter is called Landscape Cohesion (LC) (Opdam *et al.* 2003.)

landscape, and the ecological networks in it, in a functional, analytical way – perceiving landscapes from a species' perspective.

4.2   Four different assessment methods

Four different assessment methods are available with their advantages and drawbacks: (1) an empirical approach (census-based), (2) a fully mechanistic approach (PVA model-based), (3) a statistical approach (landscape-index-based) and (4) a spatial-standard-based approach (mixed). We discuss these approaches briefly here.

Much has been written recently about reserve selection algorithms based on the presence or absense of species. Pressey *et al.* (1997), for example, compare 30 different algorithms. These approaches, however, do not take into account spatial relations and are thus unsuitable for habitat networks in landscapes where metapopulation dynamics are expected to occur. Some distribution-data-based approaches do take these spatial relations into account (Hanski 1994, 1997). However, in highly fragmented landscapes, the occurrence of a species at a certain moment in time does not mean that the species has a sustainable population, for at least two reasons. First, metapopulation dynamics, i.e. local extinctions and recolonisations, limit the value of a single snapshot. In conservation planning for metapopulations of many species, it would, for example, not be a sound strategy to conserve all the patches where the species was present at a certain moment in time, and destroy the others. Moreover, what we see as distribution patterns of species is the result of historical land use patterns and historical events, and we have no certainty that populations or metapopulations are in equilibrium with the current landscape. More probably, the populations are lagging behind the landscape changes (Tilman *et al.* 1994). For both reasons, landscape conservation planning cannot be based entirely upon distribution data. Another important practical limitation is that gathering distribution data is very time-consuming and costly and often there is simply not enough time to do so. Finally, even if the data are available, it is usually only possible on the basis of the data to predict effects of habitat destruction scenarios, not of nature restoration scenarios. For a more detailed discussion of the drawbacks of the more sophisticated presence/absence-based models, see Ter Braak *et al.* (1998).

Another method of assessment is using spatially realistic Population Viability Analysis (PVA) to determine the management perspectives for certain species, usually key species, indicator species, or endangered species of specific interest (Lande 1988; Lankester *et al.* 1991; Lindenmayer and Possingham 1994, 1995). As opposed to the distribution-data-based approach, here we can take into account the dynamic processes of the populations. However, there are several problems with this approach (Verboom *et al.* 1993; Ralls and Taylor 1997). It is time-consuming to unravel the life history of species to point out the relevant parameters, and find the right values for them. Moreover, such models can hardly be calibrated and/or validated because of their stochastic nature, their long time horizon, and chance fluctuations in real metapopulations. Even if we are satisfied with the model, the outcome is often very sensitive to the very exact value of certain parameters, where even a perturbation of 1% may have a large effect upon whether or not the network is found to be sustainable: the range of uncertainty of such models is often enormous. Because it is so time-consuming, such a PVA can be performed for only one or at the most a small number of species.

Instead of using species distributions or dynamic modelling, one can evaluate the landscape by calculating landscape indices: statistics such as habitat area, number and density of patches, and various connectivity or heterogeneity measures such as fractal dimension and Shannon diversity. These techniques are nowadays readily available in GIS packages. For a recent review, see Gustafson (1998). The merits of this approach, as opposed to the approach based upon distribution patterns, are that one can calculate these statistics not only for existing landscapes, but also for future projections and other scenarios, and it is not very time-consuming. Important drawbacks are the ecological interpretation of the statistics and the lack of reference values and standards. Indeed, more habitat will be better and more connectivity will in most cases also enhance population survival, but what is the critical level?

In this chapter we discuss an approach that combines the merits of the ones above, without their drawbacks. The approach is based upon ecologically scaled landscape indices (ESLIs) (Vos *et al.* 2001), Landscape Cohesion (LC) (Opdam *et al.* 2003) and the key patch approach (Verboom *et al.* 2001).

Ecologically scaled landscape indices (ESLIs) take landscape characteristics into account as encountered by the species in the landscapes, in this way adding an ecological scaling to the landscape indices. For example, distances between patches are not recorded in metres but relative to the dispersal distance of the species under consideration. Areas are not recorded in acres or hectares but in estimated carrying capacities. These ESLIs have a greater power for predicting sustainability of populations than distribution statistics and landscape statistics alone (Vos *et al.* 2001). We developed a method in which landscapes, through a set of indices that are ecologically scaled (ESLIs), are compared to a data set of spatial standards. In this mixed approach, results of PVA simulations, analysis of empirical census data and GIS techniques are combined. The derivation of this spatial-standard-based approach is described in more detail by Verboom *et al.* (2001). In the rest of this chapter, the spatial standard based approach will be explained and illustrated.

4.3   The theory and practice of functional network analysis

4.3.1 Choosing species

A landscape does not have a function for biodiversity, but for individual species. Therefore, we must scale down from landscape level to species level (Figure 4.2).

It is important to choose the right species for an assessment. For example, one can choose species that represent different life histories (e.g. small, medium, large animals), different systematic groups (e.g. birds, mammals, insects), and/or different habitat types (e.g. forest, marshland, grassland). When

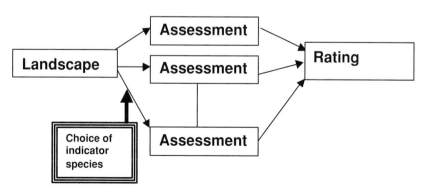

FIGURE 4.2.
Scaling down from landscape level to species-specific level and scaling up to a rating of landscape ecological spatial quality.

we are particularly interested in fragmentation, we recommend making sure the set of species contains different dispersal categories: species that disperse over small, medium and large distances, for example. A systematic way is to make a matrix of dispersal distance classes (e.g. <100 m, 100–1000 m, and 1000–10 000 m) and home range size classes, and choose representatives of various groups and various habitats to fill in the matrix. Alternatively, we can work with 'ecological profiles' (Vos *et al.* 2001) instead of true species. Or we can choose species of special interest. In the Netherlands, at the national level target species have been identified for the realisation of the National Nature Policy and the National Ecological Network (NEN). The use of true species has both an advantage and a drawback, however. We should keep in mind that we may be using assessment methods that are useful as a tool for labelling a plan with a rating, but not powerful enough for realistic predictions for particular species in particular landscapes. The choice of species should be regarded as part of the tool kit. The specific assessments are part of the tool, not the results, because their uncertainty may be high. The reader should keep this in mind when looking at Figures 4.6 and 4.7.

The scale and legend of the map, together with the scale and habitat requirements for species, also puts a limit to the species selection. For example, if a species (e.g. kingfisher) needs shallow water but the map has only one category 'water', not discerning water depth, the species is not suitable for the analysis, unless reasonable assumptions can be made about the proportion of shallow water. Another example is when a species (e.g. a butterfly) inhabits patches of one hectare or less and has a dispersal distance of less than 100 m, while the map is based on 1 km² square grids. In the latter case entire ecological networks may well lie within a single grid cell.

4.3.2 From map to functional habitat networks for metapopulations

Let us zoom in on a specific assessment for a particular species (real or imaginary) of a particular scenario (Figure 4.3).

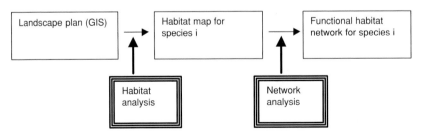

FIGURE 4.3.
A schematic representation of the derivation of a functional network map from a landscape map. This analysis is species-specific: each species has its own network.

A prerequisite for the analysis is a landscape map that has a meaningful scale and legend for the species under consideration. We need a set of rules and parameters that link the species to the map, taking into account the suitability of the map categories for the species, in terms of density and quality. For example, for a bird of old deciduous forest (e.g. nuthatch), deciduous forest of age class 80 years and more has quality 1, age class 40–80 may have quality 0.75, age class 20–40 may have quality 0.5, mixed forest aged 80+ years may have quality 0.5, old parks and gardens 0.75, etc. Now we can determine where on the map patches occur with a certain minimum quality. We can use density figures (e.g. old deciduous forest, 1 pair per hectare, etc.) to calculate carrying capacities of patches. For this analysis a database is needed with specific parameters, both species-specific and map-specific. For the Netherlands a land cover map is used that is converted into vegetation types with grid size 250 m × 250 m.

A habitat map is not enough for the determination of a meaningful rating. We need to perform a network analysis. Here again, a specific set of rules and parameters is needed. First, we merge patches that are so close together that we can consider the inhabitants together as a single interbreeding population (specific parameter: patch merging distance). Next, we omit the patches with carrying capacity of less than one pair. Last, we consider patches that lie well within the dispersal range of the species as belonging to the same functional ecological network. In merging patches and delineating networks, specific barriers prevent the formation of habitat patches or functional networks. For example, major highways dissect patches and networks for most mammals. The result of this procedure is functional habitat patches for populations and functional networks for metapopulations.

4.3.3 From functional network to specific rating

Starting with the habitat network as a basis for assessment, we need to calculate some statistics of the network, statistics that can be used as ratings. The first problem is that there are many different statistics possible, such as average carrying capacity, total carrying capacity, variance of carrying capacity, average patch connectivity, total connectivity, etc. Note that we are talking about ESLIs here, already scaled to represent the landscape from a species' perspective. The second problem is that a patch carrying capacity of 100 may be enough for a large bird or mammal, but not for an insect, so scaling ecologically is not enough. In order to get comparable results for the different species, we add an extra ecological scaling, comparing the ESLIs to specific standards for sustainability of ecological networks (Figure 4.4).

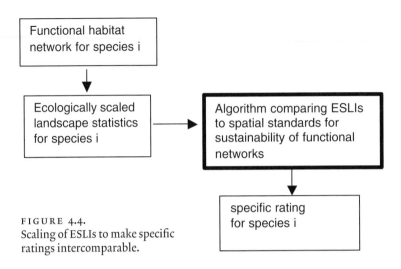

FIGURE 4.4.
Scaling of ESLIs to make specific
ratings intercomparable.

BOX 4.1. Definitions of the Minimum Viable Metapopulations approach.

A **Minimum Viable Population** (MVP) is a population that has a probability of extinction of exactly 5% in 100 years, even in complete isolation.

A **key population** is a population in a metapopulation that has a probability of extinction of less than 5% in 100 years, under the condition that immigration of at least one individual per generation takes place.

A **Key Patch** (KP) is a patch in the habitat network large enough for a key population.

A **viable population** is a population with a probability of at least 95% of surviving 100 years.

A **minimum key patch** is a patch with a carrying capacity corresponding to a probability of exactly 95% of surviving 100 years under the condition of on average one immigrant per generation.

Patch Carrying Capacity (PCC) = patch area × patch quality × average density in the patch type.

A **metapopulation** is a set of populations in a habitat network connected by inter-patch dispersal.

A **habitat network** is a set of habitat patches close enough to have a reasonable level of inter-patch dispersal.

A **Minimum Viable Metapopulation** (MVMP) is a metapopulation size or network carrying capacity corresponding to a probability of exactly 95% of surviving 100 years.

Network Carrying Capacity (NCC) = ΣPCC within network merging distance.

Network merging distance = the distance below which *c.* 80–90% of all dispersal movements take place.

For a general discussion of ESLIs, see Vos *et al.* (2001). The challenge is to link ESLIs to specific spatial standards for sustainability. For individual patches, this approach has been followed before: the MVP/MAR approach (Minimum Viable Population, Minimum Area Requirements for an MVP) (see, for example, Shaffer 1987). It consists of defining a minimum size of a population with a certain small risk of extinction, e.g. less than 5% in 100 years, and then defining the area of habitat necessary for such a population. However, in heavily human-used landscapes, this approach is not feasible, because (1) many patches are too small for MVPs for most species, and (2) patches are not completely isolated but part of an ecological network. Therefore, we derived an approach for MVMPs, Minimum Viable Metapopulations. The approach is described in full in Verboom *et al.* 2001. In Box 4.1 definitions are given; below are the results of the analysis.

We distinguish between three different levels of fragmentation, corresponding to three possible situations in which a network of patches is expected to be sustainable:

- low fragmentation: if one or more patches in the network is large enough to support an MVP (sustainable)
- medium fragmentation: if one or more patches (key patches) can support a key population. Sustainable if the other patches in the network can provide enough immigrants, i.e. if the total network carrying capacity is high enough
- high fragmentation: if no key patches occur in the network. Only sustainable if the total carrying capacity in the network is large enough to compensate for the high degree of fragmentation

Note that the minimum area requirements increase from 1 (low fragmentation) through 3 (high fragmentation). In terms of ESLIs, we calculate the carrying capacity of the largest patch (ESLI 1) and the carrying capacity of the habitat network (ESLI 2).

4.3.4   Derivation of the standards

Three sources of data have been analysed for deriving the spatial sustainability standards described above (Figure 4.5).

First, data on population turnover and presence/absence were used to derive MVP and KP standards (for a detailed description, see Verboom *et al.* 2001). Then simulations were performed with a variety of models in order to determine how much more habitat is needed for a sustainable network in fragmentation classes 1, 2 and 3 (i.e. with at least an MVP, without MVP but with at

Table 4.1. *Proposed KP (Key Patch) standards*

Species group	KP standard (reproductive units)	Range
Short-lived/small vertebrates	100	50–500
Vertebrates of intermediate size/life span	40	20–200
Long-lived/large vertebrates	20	10–100

Table 4.2. *MVMP (Minimum Viable Metapopulation) standards*

Species group	KP	MVP	Network with KP	Network without KP
Short-lived/small vertebrates	100	150	150	200
Medium-long-lived/ medium-sized vertebrates	40	60	120	200
Long-lived/large vertebrates	20	30	80	120

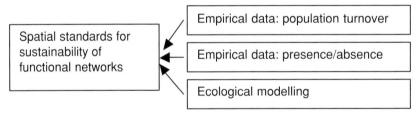

FIGURE 4.5.
Three sources of data for the spatial sustainability standards.

least a KP, with neither of these). The results of the analyses are summarised in Tables 4.1 and 4.2.

4.4 Example of a network analysis

The whole procedure described in this paper has been developed under the name LARCH (Landscape ecological Analysis and Rules for the Configuration of Habitat). We show an analysis for the Netherlands, including results of spatial policy scenarios for 2020.

Step 1: choosing species

For this particular study 18 species were chosen that are sensitive to habitat fragmentation. We present here the results for the otter (*Lutra lutra*), the nuthatch (*Sitta europaea*) and the sand lizard (*Lacerta agilis*): a mammal, a bird and a reptile, representing different habitats (marshland, forest, heathland) and different home range sizes and dispersal distances. The nuthatch is the best disperser, the lizard the worst. The otter has the largest individual area requirements, the lizard the smallest. Moreover, one species is extinct (otter), one is rare (sand lizard) and one is fairly common (nuthatch).

Step 2: habitat analysis

Figure 4.6 shows the habitat maps of the species, as derived by combining the national 250 m × 250 m square grid of vegetation data with the database in which vegetation types are linked to specific densities and qualities. Note that habitat maps are highly species-specific.

Step 3: patch definition and classification

In this step the habitat network is interpreted as patches for populations. Each fragment has a carrying capacity in the habitat map of Figure 4.6. Now patches that are separated by less than 1.5 times the radius of an individual home range, are 'merged' into patches for populations. These population-patches are then classified as too small (carrying capacity <1), MVP, large enough to contain a minimum viable population, Key Patch, too small for an MVP but large enough for a key population, or 'small patch', if the carrying capacity is greater than one but less than necessary for a key patch. Remember that the first calculated ESLI is the size (in terms of carrying capacity) of the largest patch in the ecological network. The results of this patch definition and classification step are presented in Figure 4.7. This result applies to the current landscape.

Step 4: network analysis

Next, patches that lie close enough together to form a habitat network for a metapopulation, i.e. within the specific dispersal range, are considered to form a habitat network. A habitat network is considered sustainable, and the metapopulation potentially living in it viable, if the MVMP standards are met: the network contains an MVP, or a key patch and enough accompanying habitat (see Table 4.2), or no key patch but a large network carrying capacity (ESLI 2; see Table 4.2). In Figure 4.8 the results of the network analysis are presented for the current landscape. The interpretation of the results is as follows:

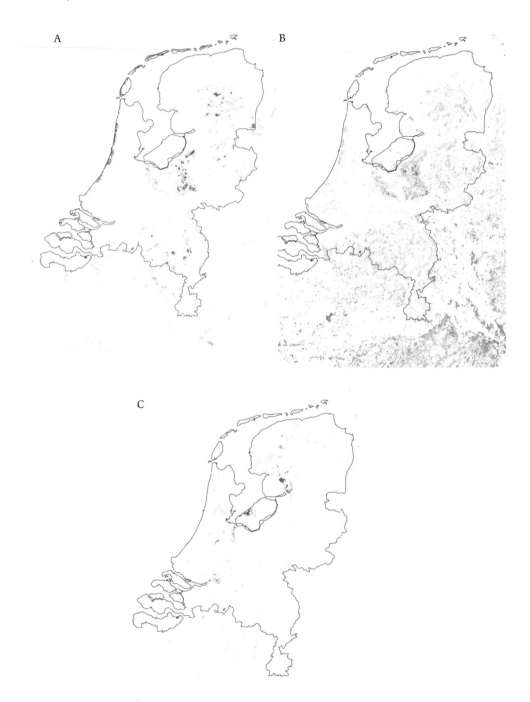

FIGURE 4.6.
Habitat of sand lizard (A), nuthatch (B) and otter (C) derived from a 250 m × 250 m vegetation map and expert knowledge of species densities in different habitat types. Darker areas indicate higher carrying capacity.

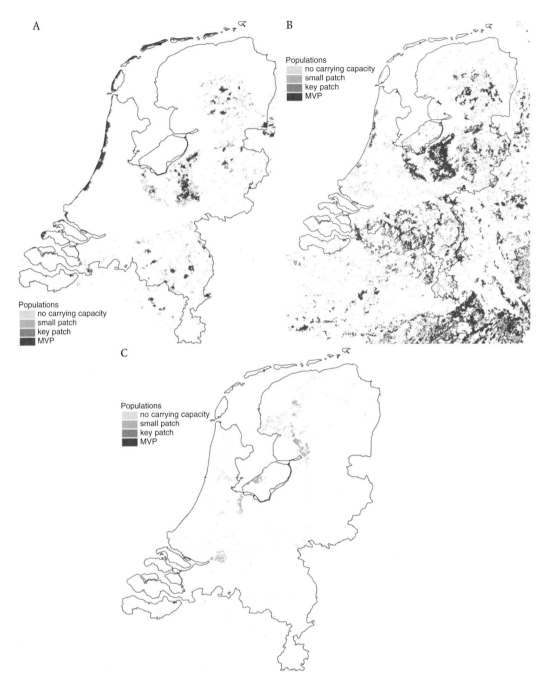

FIGURE 4.7.
Habitat classifications for the sand lizard (A), the nuthatch (B) and the otter (C) in patches with estimated carrying capacity below one pair, small patches, key patches and MVP patches.

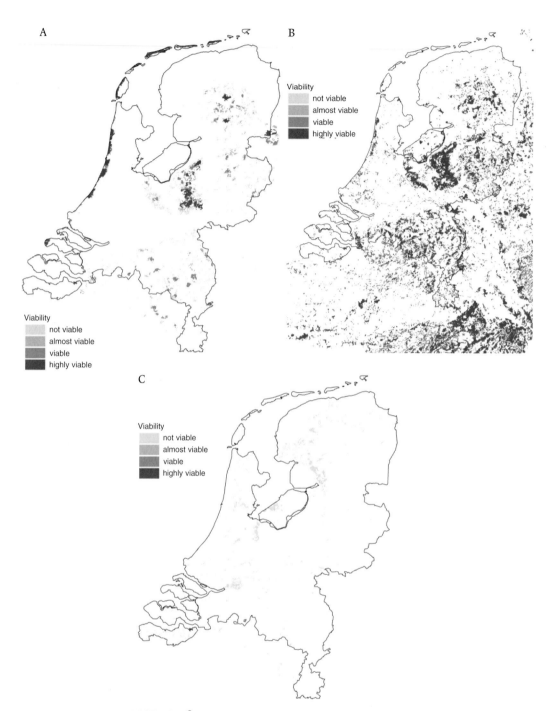

FIGURE 4.8.
Viability for the sand lizard (A), the nuthatch (B) and the otter (C), based on the
ESLI's estimated carrying capacity of the largest patch, estimated carrying capacity
of the habitat network, and the standards given in Table 4.2.

if the network carrying capacity (ESLI 2) is between 1 and 5 times what is required, the label is 'viable' (extinction probability estimated to be c. 1–5% in 100 years). If it is five times or more the label is 'highly viable' (extinction probability estimated to be <1% in 100 years). 'Almost viable' means 0.5–1 times the carrying capacity required. Note that the sand lizard (4.8A) shows a large range in viability, with some highly sustainable ecological networks (highly viable metapopulations) and some non-sustainable (non-viable) ones. The otter (4.8C) has only unsustainable habitat networks (no potential for viable populations). No key patches or MVP-size patches exist (ESLI 1) and the network carrying capacity does not meet the standards. This is partly due to the large individual area requirements of otters. The nuthatch (4.8B) has almost all habitat in highly sustainable networks, with potential for highly viable metapopulations. This is partly due to the large dispersal distance, which leads to large habitat networks for this species. Please note that the labels are applied to habitat networks, not to individual patches. So a dark colour means that a certain habitat patch belongs to a sustainable network, not that it contains a viable population. Within a habitat network, some patches may be occupied and others may be unoccupied by a species. The specific results as presented above are to be considered as a tool to reach an assessment, or rating, for a set of scenarios, and should never be interpreted as predictions.

Step 5: synthesis, from specific indices to a general rating of landscape ecological spatial quality

The outcome of the procedure can be the ratio of the carrying capacity of a network and the required carrying capacity of a network of that type for sustainability. So a value of 1.0 means that the metapopulation in the network is the minimum viable. Any network with a value above 1 is considered sustainable, and any network with a value below 1 is considered not sustainable, regardless of the species type and the network fragmentation type. This rating is specific for a network and a species. We can scale up from network level to landscape level, for example by taking the highest network value as a value for a landscape. The network values are not additive: no matter how many unsustainable (rating below 1) networks there are, the result of only unsustainable networks can never be an ecologically sustainable landscape. Neither are the ratings multiplicative, as that would mean that adding a very small, highly unsustainable network to a landscape would decrease its rating. As an alternative to the highest value, one could use the number of sustainable networks as a landscape rating for a species. Once we have obtained single values for landscapes for individual species, we can combine these into an overall rating by taking the sum of the specific ratings, or taking the number of species with at least one sustainable network.

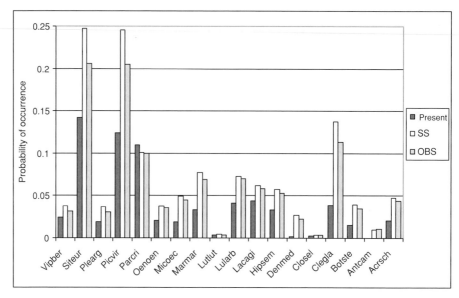

FIGURE 4.9.
Statistics for 18 species, for three scenarios. On the *y* axis is the probability of occurrence of the species in any grid cell, as calculated by combining network sustainability ratings and connectivity ratings per grid cell (Vos *et al*. 2001). The conclusion of this study is that scenario SS is the best option.

An example of a synthesis of specific results is shown in Figure 4.9. This is a synthesis from data at network level to data at the landscape level. Three scenarios are compared for 18 species. The results for three species and one scenario (current situation) have been described above and can be found here under 'Lacagi', 'Lutlut', and 'Siteur'. In this study, the probability of occurrence of species in any grid cell was estimated from the network sustainability, and the connectivity of the grid cell (see Vos *et al*. 2001).

Other synthesis options are from one species to ecosystem, by combining results for animals from e.g. forest, heathland, grassland and marshland, or from species to total biodiversity, by combining all specific results into a rating.

4.5   Discussion

We described a simple step-by-step method for assessing functional ecological networks by using metapopulation theory in terms of network sustainability for species. In this method, ecological networks are interpreted from a species perspective, and knowledge of the spatial dynamics – metapopulation dynamics – is incorporated. This method has a number of advantages and drawbacks. One of the advantages is that although it is conceptually simple, there is a whole body of scientific data and knowledge incorporated. Another advantage is the modular approach, in which all the modules can be improved, as more knowledge becomes available. See Figures 4.1–4.4 for an idea as to which parts of the method can be improved. In particular the way in which habitat patches and carrying capacities (or other relevant patch characteristics) are determined and networks are formed, and the values in the standard data set, can and should be improved in the future.

A disadvantage of the method described is the high level of uncertainty. Sensitivity analysis (Houweling *et al.* 1999) showed that the outcome could be sensitive to small alterations in the parameter values of e.g. patch merging distance, network merging distance, and parameters used for translating habitat type and area into carrying capacity. A full uncertainty analysis would have to take into account all the uncertainty in these parameters, the underlying GIS map, and the values in the spatial standard data set. Such a comprehensive uncertainty analysis has not yet been performed but the outcome can be predicted: a high level of uncertainty of the results. We think the best way to deal with this is to use the method for comparing scenarios, focusing not on the absolute values of the results, but on the main differences between the results for the various scenarios, and for species groups.

The unique property of the LARCH model is that it assesses ecological networks as functional networks, with reference to the way species are able to survive in them. This means perceiving landscapes from a species' perspective. Not the structure of the landscape as we perceive it, but the species' perspective counts. Therefore, every species has its own functional ecological network. A landscape can be suitable for species A, but too fragmented for another species B. This can be the result of (1) species B being a habitat specialist and A a generalist, e.g. A occurs in all forest and B only in old forest, (2) species B being a bad disperser and species A a good disperser, or (3) B requiring larger home ranges than A, or (4) a combination of these factors. The results for nuthatch, sand lizard and otter illustrate these differences.

Spatially realistic PVA models such as METAPHOR, VORTEX, ALEX or RAMAS-METAPOP can also be used for ecological impact assessment. When one focuses on the sustainability of an ecological network for one species of special interest, a highly endangered species, or a target species for conservation, it is better to use a detailed simulation model that takes into account all the details of this species. Such a study is usually quite time-consuming and it is therefore not feasible to perform the analysis for a large number of species and/or scenarios. Therefore, more general models (e.g. LARCH) are often preferable for general and quick comparisons of scenarios.

As a last point we would like to stress again that model assessments of all kinds, including LARCH, show high levels of uncertainty. It needs stressing that all such methods should be used only comparatively, comparing land use scenarios. The LARCH model results can be used to prioritise scenarios, not for assessing the (potential) long-term viability of metapopulations of particular species in particular locations: these tools are not meant and unsuitable for predicting the future.

5

Impacts of roads on ecological networks and integration of conservation and transportation planning: Florida as a case study

5.1   Introduction

Human population growth and land development over the last century have resulted in widespread habitat loss (Kautz 1992) and fragmentation (Harris and Silva-Lopez 1992), and increased wildlife mortality on highways (Gilbert 1996; Smith 1996). The consequences of these impacts on native species diversity can be staggering. In most cases, species more adaptable to human change in the landscape (habitat generalists, medium large carnivores, and omnivores) flourish, while habitat specialists and larger, carnivorous forms decline (Crooks and Soulé 1999; Rogers and Caro 1998; Harris and Gallagher 1989).

One approach that mitigates the negative effects of expanding human development is the creation of 'greenways' or ecological networks consisting of an integrated system of habitat corridors, core habitat reserves and naturally isolated areas. Such 'green' infrastructure can maintain or restore necessary ecological processes (e.g. animal and plant migration and dispersal, hydrologic flows, fire management regimes) to once contiguous natural landscapes (Bennett 1999; Dobson *et al.* 1998; Csuti 1991; Harris and Scheck 1991; Harris and Gallagher 1989; Eisenberg 1986; Forman 1983). Creative use of linear parks and greenways (open space) can also restore natural amenities to urban areas and provide buffers to protect surrounding natural systems (Smith and Hellmund 1993; Little 1990).

While greenways may provide a means to restore ecological function to large-scale habitat conservation systems, most are intersected by roads that act as significant barriers to ecological processes. Proper management of these greenway–highway interfaces is necessary for maintaining the integrity of an ecological network. The Florida Department of Transportation (FDOT) recognised the importance of these interfaces and funded a three-year study to identify and prioritise needs, and develop programmes for mitigation at these intersections along state-maintained roads.

This chapter will focus on the impacts of roads on greenways in the State of Florida, including (1) recent trends in population growth and infrastructure development in Florida, (2) ecological effects of roads, (3) examples of projects designed to restore landscape connectivity, and (4) approaches for integrating conservation and transportation planning.

5.2 Recent trends in population growth and infrastructure development in Florida

5.2.1 Recent population growth and infrastructure needs

Florida is now the fourth most populous state behind California, Texas and New York (BEBR 1998). Population growth between 1990 and 2000 was 24% (USCB 2000). Twenty-eight new residents move here every hour, the third highest growth rate in the USA. The population was 16 million in 2000 (USCB 2000). In addition, Florida receives an estimated 43 million tourists annually (APA 1995). Fifty per cent of the tourists visiting Florida arrive by automobile (FDOT 1994).

The impact of the large number of vehicles (12 million private registrations) (FHWA 1996) travelling on Florida's roads manifests itself through increased traffic density. Vehicle kilometres driven on State roads in Florida have increased some 90% between 1980 and 1997, nearly doubling from 194.8 million km to 372.1 million km, respectively (FDOT 1998). Florida's traffic volume is the third highest in the USA (FHWA 1996). Increased traffic densities have been shown to increase the barrier effects of roads for wildlife (Fahrig *et al.* 1995; Curatolo and Murphy 1986).

5.2.2 Highway construction

Recently, hard surface roads have been constructed at a rate of 8.8 km per day in Florida (FDOT 1998). Florida ranks tenth and thirteenth in the USA for total length of roads (182 735 km) and road density (1.31 km km^{-2}), respectively (FHWA 1996). A European comparison, the Netherlands, has 113 419 km of roads, a road density of 2.73 km km^{-2} (Veenbaas and Brandjes 1999) and a population over 15 million (Walker 1997). Urban roads in Florida have increased from a share of 16% in 1947 to 43% in 1995, whereas rural roads have declined from 84% in 1947 to 57% in 1995. In actual kilometres, both categories nearly doubled from 1964 to 1976 (Figure 5.1), although subsequently the rural category has declined (−3542 km) while urban continues to rise (+29 463 km).

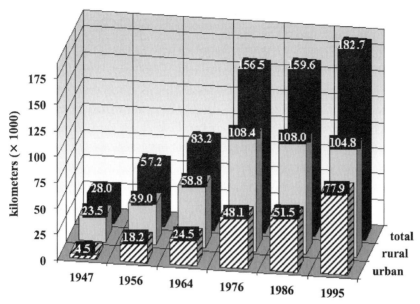

FIGURE 5.1.
Kilometres of paved roads in Florida from 1947 to 1995. Overall highway construction has increased five-fold since 1947. There are 17 times as many urban roads in 1995 as in 1947. Kilometres of rural roads have also increased since 1947, but have remained relatively constant since 1976. These trends reflect early agricultural development and later the rapid transformation of rural areas to urban land uses. Data source: Florida Statistical Abstracts 1976, 1988, 1996; Federal Highway Administration 1947, 1965.

Two causal factors are at work here. First, the figure does not reflect the amount of rural land that has been converted to urban land uses over this period, resulting in rural roads being reclassified as urban roads. Therefore the length of urban road is increasing owing to construction of new urban roads and through land use conversion. Secondly, this means that, for the length of rural road to remain relatively constant, new rural road construction is continuing at approximately the same rate that rural land is being converted to urban land uses. Urban and rural hard surface roads were constructed at rates of 4.19 and 4.64 km per day, respectively, between 1947 and 1995. Over the 48-year interval, rural roads increased by 81 300 km and urban by 73 400 km (Figure 5.1). As a result, roadless areas have decreased dramatically in size and amount (see section 5.2.3 on land use change for additional discussion).

5.2.3   Land use change

Florida has a total land area of 139 697 km², making it the twenty-sixth largest state of the USA. Although 30% of this area (38 000 km²) is protected as public parks, recreation and preserve lands, the threat of development to key areas and linkages of Florida's statewide habitat system remains. Since 1936, growth in the state has resulted in the loss of 54% of herbaceous wetlands and 22% of woodlands (Kautz 1992; Florida water management districts, unpublished data, 1995). According to Kautz (1992), Florida has lost the equivalent of eight Apalachicola National Forests since 1950, an area of approximately 2400 km². This land has largely been converted to urban and agricultural lands. From 1936 to 1994, urban lands have gone from 2800 to 19 700 km², greater than a six-fold increase over the past 58 years (Kautz 1992; Florida water management districts, unpublished data, 1995). Land conversion to agriculture had reached 44 000 km² in 1980, but now appears to be declining, a victim of urban development pressure itself. Furthermore, more than 13% of Florida has been covered by pavement (Florida water management districts, unpublished data, 1995). In summary, over 43% of Florida's original forest and wetland habitat types have been converted to human-oriented land uses (Kautz 1992).

There are now only eight remaining areas greater than one thousand square kilometres in Florida where paved roads are absent. There are 18 additional areas greater than 500 km². Combined, these areas are arranged into 12 separate contiguous regions (Figure 5.2). The largest is the greater Everglades – Big Cypress region (13 900 km²), which consists of nine adjacent habitat blocks separated by major paved roads. Seven of the twelve regions consist of public forest, park, preserve, wildlife refuge or military lands; the remaining five are in part targeted for public acquisition. The state resembles a jigsaw puzzle divided into various sized pieces of land separated by an extensive road network. This configuration jeopardises the ability of wide-ranging species such as the Florida panther *Felis concolor coryi*, Florida black bear *Ursus americanus floridanus* and river otter *Lutra canadensis* to successfully disperse and colonise adjacent habitat areas.

5.3   Ecological effects of roads

Pronounced ecological effects occur at the interface between highways and habitats. These effects include wildlife–vehicle collisions, spread of invasive species, barriers to wildlife movement, habitat degradation from pollutants, habitat loss and fragmentation, adjacent habitat modifications, and negative edge effects (Stiles and Jones 1998; Suarez *et al.* 1997; Marini *et al.* 1995; Cristoffer 1991; Harris and Scheck 1991; Southall 1991; Wilcox 1989; Verkaar 1988; Harris 1984; Mader 1984; Muskett and Jones 1980; Leedy 1975). Of primary importance associated with highway–greenway interfaces are barrier effects of the road itself and habitat fragmentation. Although collisions between vehicles and wildlife represent the obvious more graphic effects of roads

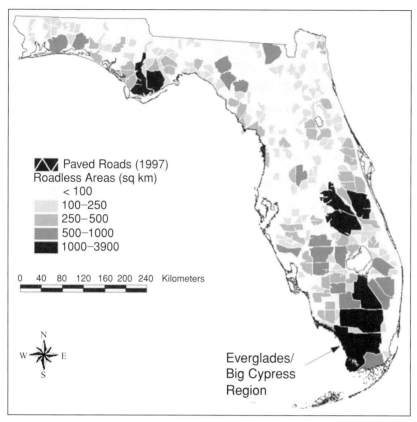

FIGURE 5.2.
Roadless areas in Florida (includes paved roads only). Only three regions remain that have areas greater than 1000 km². (Data from Florida Department of Transportation, 1997.)

on wildlife, fragmentation of contiguous habitat can result in more pervasive impacts including loss of populations and species.

5.3.1 Highway mortality

The Florida panther and Florida black bear present acute accounts of the impact of highway mortality on populations of large threatened and endangered species. Automobile-related deaths of the critically endangered Florida panther gained worldwide attention in the 1980s with the planned extension and expansion of I-75 (Alligator Alley). Between 1979 and 1989 an average of 1.7 panthers were killed per year on highways in south Florida. Several studies discuss issues associated with the highway project on Florida panthers in south Florida, including installation and effectiveness of wildlife underpasses (Logan and Evink 1985; Foster and Humphrey 1995; LoBuono 1988).

Recent road improvement and new road construction projects have heightened public awareness about auto collisions as a major human-related cause of black bear mortality. From 1976 to 1999, 729 black bears died from vehicle collisions on Florida highways (FFWCC 2000). Annual mortality totals from vehicle collisions continue to grow every year. In 1998, 88 black bear deaths were recorded on Florida highways, a 19% increase from the previous year and 2.4 times that documented 10 years earlier (FFWCC 2000).

The threat is even greater when you consider the exceedingly low population levels. There are an estimated 46–74 Florida panthers (Maehr 1997a), and only seven core populations of Florida black bears with some 2000–2500 individuals (Eason and O'Meara 2000; W. McCown, FFWCC, pers. comm.).

5.3.2 Habitat loss, fragmentation and isolation

Habitat isolation and fragmentation by linear structures such as roads may have drastic effects on biological diversity. Human influence threatens native biological diversity through loss of species from genetic inbreeding, elimination of large uninterrupted habitat, and invasion of alien species (Forman and Alexander 1998; Andrews 1990; Harris and Gallagher 1989). The rapid fragmentation of the landscape by roads and urbanisation has resulted in the loss of normal dispersion patterns (Harris and Scheck 1991; Andrews 1990). Landscape character has switched from humans in a natural landscape matrix to one of natural areas in a human-dominated landscape. The erosion of native biodiversity in Florida is rapidly affecting everything from individual genetic variability to landscapes (Harris and Scheck 1991).

Maehr and Cox (1995) demonstrated the importance of habitat contiguity for the Florida panther in South Florida. By correlating landscape features with radiotelemetry point locations, they were able to show that panthers preferred large forested habitat areas and avoided open cover types. Size of occupied suitable habitat patches averaged 20 816 ha (Maehr and Cox 1995). Additional analysis showed that only six adequate roadless areas of this size remain in Florida. Four of these areas, all located in south Florida, are contiguous, but separated by major highways (see Figure 5.2). Furthermore, collisions with automobiles represent the principal means of human-related mortality for the Florida panther (Maehr et al. 1991). Successful dispersal and expansion of the current range of the Florida panther will depend on provision of large contiguous forested habitat areas.

Likelihood of patch occupancy by black bears in Florida appears to increase as patch size increases (Hellgren and Maehr 1992). In fact, bears utilised public preserves of less than 100 000 ha only when adjacent to larger occupied tracts of land (Hellgren and Maehr 1992). Key management issues for Florida panther and black bear include increasing permeability of roads, and reducing habitat loss and fragmentation through creation of habitat networks and cooperative land management schemes with private landowners.

5.4 Projects for restoring landscape connectivity

Many methods and devices have been tested and used to help reconnect landscapes and prevent wildlife–vehicle collisions. Some examples include wildlife underpasses, tunnels and ecoducts, barrier fencing, creation of new habitat areas, headlight reflectors, sound devices and infrared sensors for cars. These methods and devices have been utilised with varying effectiveness (Evink *et al.* 1996, 1998, 1999; Jackson 1996; Madsen 1996; Langton 1989; Hunt *et al.* 1987; Ford 1980). Issues such as proper vs. improper design, proper vs. improper placement at heavily used crossing sites or good maintenance on fencing or barriers that contain the animals vs. poor maintenance that allows animals to circumvent the device can be keys to success or failure. In Florida, the first measures taken to reduce roadkills and to improve landscape connectivity occurred in the 1980s.

5.4.1 Alligator Alley (I-75) and the Florida panther

Alligator Alley (Interstate 75), a 125 km highway traversing the Big Cypress Swamp and Florida Everglades, was a four-lane motorway in the 1980s. The Florida panther recovery plan considered road mitigation necessary for the restoration of a viable population in south Florida (USFWS 1995). Specific prescriptions included wildlife underpasses, warning signs, reflectors, and nighttime speed restrictions. The FDOT installed 24 underpasses designed and located specifically to facilitate safe travel for the Florida panther (Villano 1993). Figure 5.3 shows the design of the underpasses constructed on I-75/Alligator Alley.

These modifications resulted in the elimination of vehicle collisions with the endangered cats on I-75 (G. L. Evink, pers. comm.). Noted as the reason for the effectiveness of the underpasses was the amount of preplanning and research for location and design of the structures. Openness of the structure (openness ratio > 0.92), barrier fencing 3 m high that runs the length of the project boundary (64 km) and native vegetation were critical aspects of the underpass design (Foster and Humphrey 1995; Evink 1990). The success of the I-75 underpasses has led to construction of six additional wildlife crossing tunnels on SR 29, a north–south arterial highway that also crosses primary habitat for the Florida panther.

FIGURE 5.3.
One of the functional underpasses constructed on I-75/Alligator Alley in the Big
Cypress National Preserve in south Florida. Bridge dimensions are 36.6 m wide ×
2.4 m high (Evink 1990). Photo credit: D. Maehr, University of Kentucky, USA.

5.4.2   Central Florida and the Florida black bear

The pattern of intensified development and road construction in Florida
has produced a clumped distribution of the threatened Florida black bear,
segregated into several subpopulations (Figure 5.4). An area of great concern
for black bear conservation is preserving the landscape linkages between the
Wekiva River basin, north of Orlando, and the Okefenokee Swamp National
Wildlife Refuge in southern Georgia (Figure 5.4). These areas are separated by
approximately 250 km and encompass over 750 000 km^2 of existing and proposed conservation lands. Twenty-one major roads pose serious threats to the
integrity of this network, not only as barriers to movement but also as development facilitators that result in habitat loss and fragmentation. Two hundred
and seventy-five black bear roadkills were recorded on these roads from 1976
to 1999 (FFWCC 2000).

Five of these roads, located within the Ocala National Forest or the Wekiva
River basin (Figure 5.5), account for the majority of these deaths: SR 46 (48),
SR 40 (69), SR 19 (58), SR 44 (26) and CR 42 (21). To answer questions on bear
movements in this area, in 1994 the FDOT constructed an experimental con-
crete culvert 2.4 m × 7.3 m (Figure 5.6) under SR 46 (Roof and Wooding 1996).
In addition, a 3 m high barrier fence extending 0.6 km west and 1.1 km east

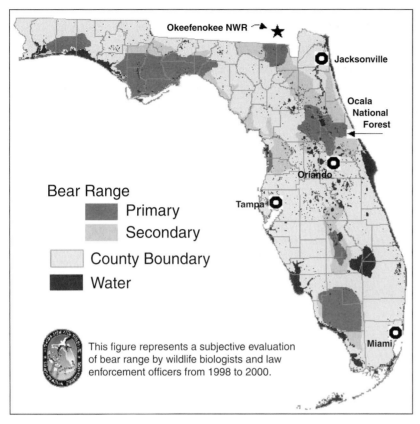

FIGURE 5.4.
Current distribution of the Florida black bear. Habitat loss and fragmentation have
reduced the former statewide range of the black bear to seven areas of concentration.
Map source: Florida Fish & Wildlife Conservation Commission.

of the underpass was erected (Roof and Wooding 1996). The aim of the project
was to reduce the number of roadkills occurring along this 4.5 km section of SR
46 (20 fatalities from 1981 to 1994) (Gilbert and Wooding 1994). Construction
of the culvert was also intended to maintain a connection between the isolated
Wekiva population and the larger Ocala National Forest population of over 500
individual black bears (Gilbert and Wooding 1994; Eason 2000). Construction
of additional underpasses and barrier fencing is planned to effectively retrofit
the remainder of this section of highway to reduce future roadkills and increase
permeability.

Culverts and underpasses serve as connections between landscapes divided by highways and play a critical role in decreasing the barrier effect of
roadways by increasing permeability for wildlife. Increased permeability results in consequent decreases in mortality (Yanes *et al.* 1995). These underpasses can facilitate corridors that connect spatially separated habitats and

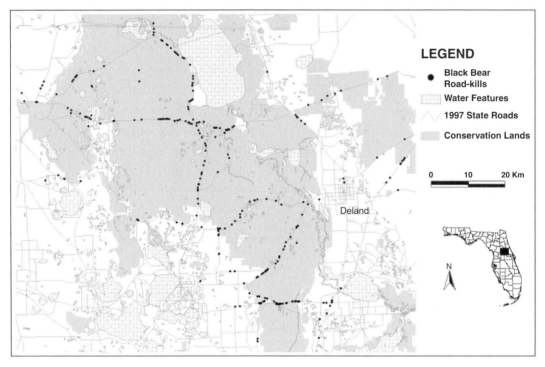

FIGURE 5.5.
Ocala National Forest region including location of black bear roadkills on major highways in central Florida. From 1976 to 1999 over 200 roadkills have been documented on these five major roads. Data source: Florida Fish & Wildlife Conservation Commission, 1999.

enhance the efficacy of wildlife movement throughout the landscape (Forman 1983).

5.5 Approaches for integrating conservation and transportation planning

In Goal 5 of the 2020 Florida Transportation Plan (FDOT 1995), 'Travel choices to ensure mobility, sustain the quality of the environment, preserve community values, and reduce energy consumption', the FDOT seeks strategies to reduce urban sprawl and encourage transit use. Conservation strategies include support for ecosystem management through mitigation banking and greenway planning, and increased use of native planting on rights-of-way (FDOT 1994). Two projects were sponsored by the FDOT that apply landscape ecological principles to infrastructure planning: (1) The Florida Statewide Greenways Plan, and (2) Identification and Prioritisation of Highway–Greenway Interfaces.

FIGURE 5.6.
Precast concrete culvert constructed on SR 46 between the Wekiva River State Park and Seminole State Forest in central Florida. Placement coincides with high incidence of black bear roadkills (16 from 1981 to 1998). Dimensions are 2.4 m high × 7.3 m wide × 14.6 m long. Data source: Florida Fish & Wildlife Conservation Commission, 1999.

5.5.1 Planning for habitat connectivity

Along with the encroachment of roads and development, isolation and fragmentation of ecosystems and associated wildlife populations has occurred, with varying detrimental effects. One solution to this phenomenon was presented by Harris (1985) as 'conservation corridors' or landscape linkages to connect large conservation areas. The potential positive and negative aspects of this approach to landscape reintegration have since been discussed by numerous conservation biologists (see Bennett 1999; Beier and Noss 1998; Dobson *et al.* 1998; Csuti 1991; Soulé 1991a; Noss 1987b; Simberloff and Cox 1987; Noss and Harris 1986).

Land bridges connecting continents, according to palaeontological records, provide evidence of use of corridors by entire faunas (Harris 1985). Physical connectors of sufficient size to promote genetic flow in two directions of full and balanced plant and animal communities are necessary to maintain intended ecosystem functions and processes (Noss and Cooperrider 1994; Noss 1991; Harris 1985). Standards below this objective may serve only to increase the spread of 'weedy species' and maintain the isolation of interior specialists and those requiring large territorial boundaries (Harris 1985; see also Beier and Noss 1998; Dobson *et al.* 1998; Simberloff and Cox 1987). Eisenberg (1986) explains that these linkages may not increase population size, but would increase the likelihood that recolonisation of locally extinct species could occur. In addition, such connections would protect against the weakening effects of inbreeding (Eisenberg 1986).

These corridors must have sufficient width to maintain interior habitat qualities that would enhance use by threatened area-sensitive species (Noss 1983; see also Noss and Cooperrider 1994; Soulé 1991b). Design of these landscape connections, however, is thwarted by the presence of roads within and surrounding habitat islands. These roads act as barriers to dispersal and function as 'wildlife killing machines' (a term coined by Walt Thomson of the Florida Department of Environmental Protection, in Harris and Gallagher 1989). Studies of reserve network design recognise road density as a critical indicator for evaluating system integrity (Noss 1995). Roads, as a barrier to animal movement, are considered one of the six major determinants of functional connectivity (Noss and Cooperrider 1994). Recommended designs now illustrate use of wildlife crossings to permeate transport facilities (Noss 1995).

The Florida Statewide Greenways Network Plan was designed to provide guidance for conserving valuable natural resources of Florida and to restore connectivity between core conservation reserves and other isolated conservation areas (Figure 5.7). As defined by the University of Florida Greenways Planning Team (FGPT), 'a greenway is a corridor of protected open space that is managed for conservation and/or recreation' (FGP 1999). They function as linkages between parks and nature reserves to create an interconnected system. The FGPT followed a regional landscape approach designed to establish a plan for an ecologically functional statewide habitat system (Hoctor *et al.* 2000). The recommended greenways plan includes 9.31 million hectares (57% of the state). Of this area 63% is public land, CARL or SOR proposals, or open water (FGP 1999). These lands include many large natural hubs (core conservation areas) and connecting linkages (Figure 5.7). See Chapter 13 for a more detailed account of the Florida Greenways Project analysis rationale and results. Within the core conservation areas and associated linkages identified in the greenways plan are many intersections with roads. As part of the design of a functionally integrated ecological network, many of these road intersections will require some type of mitigative measures such as wildlife underpasses.

The use of highway crossing structures at intersections with habitat linkages or corridors offers a method to reduce transportation-related wildlife mortality and restore connectivity to the landscape. It has provided ecologists and engineers with an opportunity to reduce the negative effects of roads, by

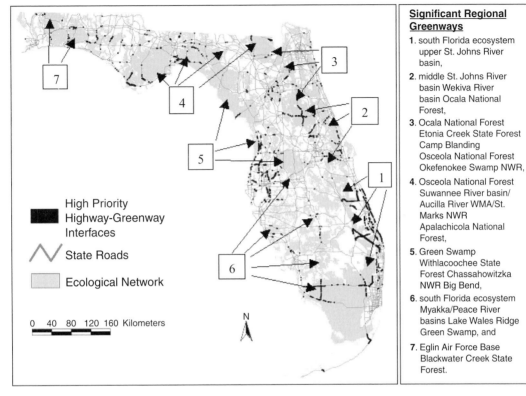

Significant Regional Greenways

1. south Florida ecosystem upper St. Johns River basin,
2. middle St. Johns River basin Wekiva River basin Ocala National Forest,
3. Ocala National Forest Etonia Creek State Forest Camp Blanding Osceola National Forest Okefenokee Swamp NWR,
4. Osceola National Forest Suwannee River basin/ Aucilla River WMA/St. Marks NWR Apalachicola National Forest,
5. Green Swamp Withlacoochee State Forest Chassahowitzka NWR Big Bend,
6. south Florida ecosystem Myakka/Peace River basins Lake Wales Ridge Green Swamp, and
7. Eglin Air Force Base Blackwater Creek State Forest.

High Priority Highway-Greenway Interfaces

State Roads

Ecological Network

0   40   80   120 160 Kilometers

N

FIGURE 5.7.
Identified high priority greenway–highway interfaces. Major ecological greenways are numbered and described in the key. The map displays important linkages necessary to maintain connections between remaining primary habitat areas for wildlife populations including the Florida panther and black bear. Road segments that intersect greenway linkages were prioritised according to various ecological criteria and are shown in black. Many of these may act as barriers to wildlife movement and alter various landscape processes.

restoring natural processes as they occurred prior to fragmentation of the landscape (for example, wildlife movement and migration, flood, and fire).

5.5.2 Identifying landscape linkage barriers

Governmental efforts and public support in the 1990s toward establishment of greenways and ecologically based landscape linkages across Florida prompted FDOT to look at highway–greenway interfaces, and the potential for implementing an underpass construction programme of regional and statewide scope designed to restore landscape connectivity and ecological processes. The FDOT initiated the research project described here to provide a framework for integrating road projects with the promotion of the greenways and other state conservation programmes such as Preservation 2000/Florida Forever 2010

(Florida Forever Advisory Council 2001) and Florida GAP analysis studies. This research included the development of an algorithm for analysing existing datasets to identify and prioritise highway–greenway interfaces to evaluate the need for wildlife crossing structures or underpasses (see Smith 1999b).

Important environmental factors for prioritising relative impact of roads on conservation value were established by conducting a survey at the FDOT sponsored 'Transportation Related Wildlife Mortality Seminar' in Orlando, Florida, in 1996. Fifty-three ecologists/biologists and twelve engineers/planners filled out a questionnaire. Respondents were asked to list and rank criteria considered important for prioritising locations for wildlife underpasses to alleviate roadkills and to restore ecological linkages. The original eleven elements identified and ranked by the respondents are listed below.

1. Chronic roadkill sites
2. Known migration or movement routes
3. Identified hot spots of focal species
4. Landscape linkages (designated greenways)
5. Presence of listed species
6. Identified strategic habitat conservation areas (SHCAs)
7. Riparian corridors (with potential for retrofitting existing structures)
8. Core conservation areas
9. Presence of separated required ecological resources (e.g. a forest patch and ephemeral wetland breeding area for amphibians separated by a highway) for a species or set of species
10. Public ownership (or in public land acquisition programme) as opposed to private lands
11. Potential to be included in proposed road improvement project

These were fitted to ten corresponding GIS data layers (see Table 5.1).

1. Chronic roadkill sites of listed species
2. Predicted wildlife movement routes based on:
   • topography: ridge–valley configuration
   • hydrography: riparian corridors
   • habitat/land cover type
3. Hot spots of focal species
4. Landscape linkages, a.k.a. 'greenways'
5. Presence of listed species
6. Strategic habitat conservation areas
7. Public conservation lands
8. Proposed road improvement projects

Table 5.1. *Grid values for highway–greenway interface priority model*

Abbreviations: Clands98, State-designated conservation lands; SHCA, Florida Fish and Wildlife Conservation Commission designated strategic habitat conservation areas; FNAI, Florida Natural Areas Inventory (1997) (priority designations a,b,c); TNC, The Nature Conservancy (charrette priorities).

Category	Criteria	Base	Survey	Multiplier
Landscape features	Gradients		2	9
	Topography: ridges (greater than 36 m)	2		
	Ecotone (natural lands greater than 40 ha)	2		
	Riparian		7	4
	Streams/lakes in natural habitats	4		
	Canals in natural habitats	3		
	Streams/lakes/canals in urban/agriculture	2		
	GFC habitat/land cover		not included	3
	Xeric habitats	4		
	Wetland habitats/hardwood hammocks	3		
	Silvicultural/mixed pine and hardwoods	2		
Biological features	GFC hotspots of focal species		3	8
	7+ species	4		
	5–6 species	3		
	3–4 species	2		
	Element occurrences (listed species locations)		5	5
	Endangered	4		
	Threatened	3		
	Species of special concern/bird rookery	2		
Roadkill	Roadkill		1	10
	Listed species (black bear/panther/key deer)	4		
	State Parks	2		
Planning	Strategic Habitat Conservation Areas		5	5
	High (Clands98(proposed), SHCA(proposed), FNAI(A,B), TNC(Priority))	4		
	Low (FNAI(C), TNC(Interest))	2		
	Greenway final rankings (linkages)		4	7
	High priority	4		
	Medium priority	3		
	Low priority	2		
Public	Public Lands		10 (*8)	2
	Clands98 (existing)	4	*	
Infrastructure	Road projects		11	1
	Proposed, bridge replacements	4		
	Existing	2		

* Includes core conservation areas.

Attributes of each dataset were assigned base values for analysis. For example, the 'listed species' dataset was broken down into three classes with descending values of importance: endangered = 4, threatened = 3, species of special concern = 2. These datasets were then grouped into six categories (biological features, landscape features, roadkills, planning, infrastructure and public conservation lands) to control for redundancy in the information. Table 5.1 displays the categories, the criteria and associated attributes with assigned base values. The survey rank and multipliers reflect priority rankings from the survey questionnaire. After grouping the datasets into categories, multipliers used in the analysis were determined by averaging the values (rounded up) of all individual criteria in each group. The final multipliers are shown below.

- biological features 7
- landscape features 6
- infrastructure 1
- public lands 3 (includes ranking of core conservation areas, multiplier = 3)
- planning 5
- road-kill 9

The resolution (cell size) used in the analysis was 100 m. State roads were buffered on each side by 600 m to define an area for analysis and to account for negative edge effects (based on findings of general literature review; see Smith 1999b) of highways on adjacent habitat quality. This road buffer zone was used as a data mask for the analysis; any characteristics lying outside the buffer were not included in the priority analysis.

Cell-based modelling (ESRI, Inc., *Arcview* Spatial Analyst) was used to analyse and combine datasets. The procedure calculated a new dataset containing all possible combinations of the individual criteria (all values) for each group. The combined datasets were then processed by an automated algorithm (created with *Arcview*, Avenue) that sorted the combinations, applied the user-input weightings, and summed the weighted data values for each 100 m × 100 m cell to develop a final priority layer, which consisted of a graduated scale of values ranging from 84 to 320. Priorities were assigned by dividing the total score by equal 20 unit intervals, the highest 20 values = 1, second highest 20 values = 2, third highest 20 values = 3, etc. A cell was considered to be high for a certain data category (i.e. biological features, landscape features, etc.) if it scored from one to four.

A field inventory was conducted to verify the accuracy of the GIS data and the model to prioritise or rank sites. This 'ground-truthing' was used to identify several features at each site and served as a means to characterise highly ranked highway–greenway interface zones identified by the model

(Schaefer and Smith 2000). These features included presence of existing structures (bridges, culverts, etc.), their dimensions and composition, roadway characteristics (width, number of lanes, width of paved surface), description of surrounding landscape features, identity of associated aquatic features, and signs of present animal use. GPS points were recorded and surveys performed at more than 1400 priority sites throughout the State.

The priorities determined by the model indicate significant focus toward nationally and regionally significant conservation areas and riparian corridors. Major regional greenways within the State that are intersected by high-priority road segments are shown in Figure 5.7 (see p. 85). Of a total of 15644 prioritised road segments, 12828 (82%) were associated with identified greenways, and 4019 (26%) of these were located within existing conservation lands. On proposed conservation lands there were 2469 (16%) prioritised road segments recorded.

South and central Florida

Of the highway–greenway interfaces found in south Florida, those with the greatest significance were associated with one major ecological hub (Everglades National Park/Big Cypress National Preserve) and several connecting linkages (Figure 5.7, greenway nos. 1 and 6). Field inventories were conducted on twenty-five high-priority sites on ten different highways identified in the model results and located either within the Everglades National Park/Big Cypress National Preserve or crossing the associated linkages that connect it to the Lake Wales Ridge and several riparian corridors to the north.

The most significant highway–greenway interfaces identified in central Florida were associated with two major ecological hubs (Ocala National Forest and the Green Swamp) and several major linkages including Wekiva River State Park, middle St. Johns River and tributaries, Withlacoochee State Forest, and Chassahowitzka National Wildlife Refuge (Figure 5.7, greenway nos. 2 and 5). Forty-eight high-priority sites on 26 different highways located either within the Ocala National Forest or the Green Swamp or crossing various connecting linkages to the north and south were identified and field-inventoried.

Northeast Florida and the Panhandle

The most significant highway–greenway interfaces identified in northeast Florida were associated with one major ecological hub (Osceola National Forest) and major linkages to the south, Camp Blanding Military Training Site and Etonia Creek State Forest (Figure 5.7, greenway no. 3). Fourteen high-priority sites on ten different highways located either within the Osceola National Forest or crossing the associated linkages that connect it to the Ocala National Forest were identified and field-inventoried.

The major highway–greenway interfaces identified in panhandle Florida were associated with two major ecological hubs (Apalachicola National Forest and Eglin Air Force Base) and adjacent linkages. Thirty-one high-priority sites on five different highways were identified within the Apalachicola National Forest, St. Marks National Wildlife Refuge, Aucilla River Wildlife Management Area or the Suwannee River or crossing linkages that connect them (Figure 5.7, greenway no. 4). In addition, five high-priority sections were identified on four different highways that separate and fragment Eglin Air Force Base and abutting Blackwater Creek State Forest (Figure 5.7, greenway no. 7). Ecological characterisations of all these sites were conducted during field surveys.

The field information gathered from these sites was used to give recommendations on the type of mitigation, if any, that may be necessary to alleviate the associated ecological conflict (e.g. roadkills, restriction of natural stream and floodplain dynamics, etc.). Recommendations included anything from installation or expansion of underpasses or culverts to minor measures such as fencing, signage, or speed restriction (Schaefer and Smith 2000).

For the purposes of this chapter, the most significant issues to report in regard to the actual results of the study are the effects of the contributing criteria. Many of the criteria used in this model (i.e. riparian linkages, core areas of conservation, existing and proposed public conservation lands, SHCAs, and land cover/land use) were also used for the Florida Statewide Greenways Network Planning Project. As a result, areas identified in the Florida greenways analysis were also identified as high priorities in this analysis.

Riparian systems are represented in four individual criteria used in the analysis: riparian, habitat/land cover, SHCAs, and greenway linkages. As is the case with the greenway criterion, riparian systems are an integral part of Florida's statewide conservation system. They act as refuges and travel corridors, and provide sources of food and shelter for various birds, mammals, herpeto-fauna, and fish (Schaefer and Brown 1992; Spackman and Hughes 1995; Noss 1993; Smith 1993; van Zadelhoff and Lammers 1995; Dodd 1990; Darveau *et al.* 1995). In many cases the greenway linkages contain riparian networks. It is where roads intersect these that functional terrestrial underpasses are necessary.

Biodiversity hot spots played a major role in final rankings of road segments. Hot spots that were identified by the FFWCC, in combination with listed species element occurrences, provide indications of high-priority conservation areas. The hot spots dataset, based on the presence of 44 focal species, also acts as the basis for the FWC-SHCAs used in this analysis. Presence of hot spots and/or inclusion in SHCAs coincides with the location of most major core conservation areas. Thus, most of these locations are either in public ownership or on proposed conservation lands.

Black bear roadkills significantly influenced the results at the statewide level. This was due to the high weighting that the criteria received in the analysis and the wide distribution of the species across the State. When roadkill sites are clustered or contain multiple kills, they require extra consideration regarding their potential importance as travel routes for listed species. The quality of the existing roadkill data was good regarding the three listed species available (Florida black bear, Florida panther, and Florida key deer); however, better documentation on other species would be valuable towards improving the model results.

Sixty-six state road projects were identified within highly ranked highway–greenway interface zones. These projects are scheduled through 2004 and include 11 bridge replacements/construction, 49 road expansions/ reconstruction, and 6 new roads. A key to mitigation of barrier effects of highways and automobile traffic on wildlife movement through ecological networks is the programming of road projects and identification of existing structures. Through these opportunities construction of wildlife underpasses or other mitigation measures can be programmed into the proposed road project. Such preplanning can reduce costs incurred when engineers must retrofit existing roads.

5.6   Conclusions

5.6.1 Recommended highway designs

Several factors should be considered in the design of roadways and wildlife crossing structures. Topographic relief can be used to enhance design and placement of underpasses that follow ecological flows and landscape patterns. Proper management of rights-of-way and adjacent vegetation is an important factor in the design of wildlife crossing sites (Clevenger and Waltho 2000; Hewitt *et al.* 1998; Tewes and Blanton 1998; Friedman 1997; Land and Lotz 1996; Roof and Wooding 1996; Singer and Doherty 1985). Right-of-way and road corridor design should include sufficient clearance distance from the roadway and reduction in severity of curves and slopes of hills to increase driver visibility. Appropriate native vegetation at entry points to underpasses provides cover for species intolerant of open areas and safety to species susceptible to ambush predators. Clevenger and Waltho (2000) identified proximity to human activity or influence as the most critical factor in determining species performance ratios at wildlife underpasses in Banff National Park, Alberta, Canada.

Terrestrial connections along river corridors are essential at road intersections to provide connectivity for terrestrial vertebrates to move between conservation areas. As such it is imperative that bridge replacements be pro-grammed to include accommodations for terrestrial connections adjacent and parallel to the watercourse. These connections should include native vegeta-tion consistent with the present community type.

In Florida, it has been found that functional underpasses already exist at several high-priority highway–greenway interface zones. Many of these re-quire only minor directional fencing or vegetative plantings to enhance use by wildlife and to provide connectivity to adjacent areas. Human use of these ex-isting structures should be restricted, to increase the potential use by wildlife. Examples include the existence of wide floodplain bridges constructed at stream intersections, and abandoned railway bridges through existing conser-vation areas.

5.6.2 Land ownership, conservation easements and highway–greenway interface zones

A mitigation strategy should include not only construction of under-passes or other means designed to reduce fragmentation of habitat linkages, but land use controls or conservation designations as well. Perhaps, if these areas were designated as part of a statewide habitat conservation network, land development could be directed into other less critical areas and prevent urban sprawl and strip development along roads in rural areas.

As property values along roads soar it has become increasingly difficult to justify outright purchase of land (fee-simple acquisition) to further conserva-tion goals. Opposite from outright purchase are zoning and land use restric-tions; however, these are unpopular with landowners and subject to change under political transitions, and are therefore less effective than perpetual land conservation devices (Wright 1994).

Alternative strategies that are becoming more popular include control of partial interests of land such as conservation easements through donated de-velopment rights (DDRs), purchased development rights (PDRs), and transfer-ence of development rights (Wiebe 1997). In Florida, PDRs have been used to establish conservation easements on commercial forest lands in critical flood-plain management and aquifer recharge areas.

Over 57% of the State of Florida has been identified as part of the Florida statewide ecological network (FGP 1999). Of those lands identified, 22% are in private ownership with little or no development restriction. The use of DDRs and PDRs can play a critical role in protecting these private lands from devel-opment, aside from costly fee-simple acquisition purchases that remove land from property tax rolls. Considering that conservation easements were deemed appropriate for protecting scenic vistas in the 1930s (Haapoja 1994), they are just as essential now for maintaining critical linkages at highway interfaces for an ecological network.

5.6.3 Planning and research to integrate transportation and conservation objectives in Florida

Co-ordination of the results of the highway–greenway interface prioritisation model with those of the Florida statewide greenways model is necessary to target road sites for addressing mitigation needs to promote furtherance of goals to conserve Florida's remaining biodiversity. Accelerated facilitation of this goal can be accomplished by using the model results to program mitigation with future road projects and by using the inventory of existing infrastructure (bridges and culverts) at prioritised greenway–highway interface zones to evaluate their potential to improve permeability of roads. Another major effort under way is a two-year project to monitor use of existing bridges and culverts by various wildlife species. Information generated by this study will be used to develop design standards for new wildlife crossing structures that identify preferences by taxa within various ecosystem contexts.

IRENE M. BOUWMA, RUUD P. B. FOPPEN
AND ALEXANDER J. F. M. VAN OPSTAL

6

Ecological corridors on a European scale: a typology and identification of target species

6.1 Introduction

The ongoing fragmentation of natural and semi-natural areas in Europe and the development of related concepts in population dynamics and landscape ecology in recent decades has raised awareness of the role that connectivity and permeability of landscapes play in maintaining viable natural populations. Since the publication of the 'island theory' by MacArthur and Wilson (1967) several studies have shown that populations are affected by the spatial configuration of their habitat. The metapopulation concept has been accepted by landscape ecologists as an adequate way to describe the relationship between size and connectivity of habitat patches and population persistence.

One of the questions regarding permeability of the landscape for species is whether species require specific landscape structures to move from one suitable habitat patch to another. Within the ecological network concept these 'structures' are referred to as corridors. Within ecological networks ecological corridors are various landscape structures, other than core areas, in size and shape varying from wide to narrow and from meandering to straight, which represent links that permeate the landscape, maintaining or re-establishing natural connectivity (Jongman and Troumbis 1995). Within an ecological network, corridors could be multi-functional landscape structures.

In Europe, ecological corridors are often the result of human intervention in nature: hedgerows, stone walls, landscapes with small forests, canals and regulated rivers. The nature of ecological corridors and their efficiency in interconnecting remnants and in permeating the landscape depend on the habitat site they originate from and the land use mosaic within which they are embedded

and of which they consist (Forman 1983). Their density and spatial arrangement change according to the type of land use. Their connectivity varies from
high to low depending on their spatial arrangement, internal structure and
management (Jongman and Kristiansen 2001).

However, a standardised definition of corridors is lacking. Since 1995 the
development of the Pan-European Ecological Network has become a priority in nature conservation policy in Europe (see Bennett 1998; van Opstal
1999). An important aspect of the Pan-European Ecological Network is that
it ensures the ecological coherence within Europe as a whole, by ensuring
sufficient opportunities for the dispersal and migration of species and habitat patches of sufficient size to ensure a favourable conservation status for
species.

For the development of the Pan-European Ecological Network there is a need
to clarify the terminology on ecological corridors as well as to review the application of ecological corridors on a European scale. Often corridors are designed
without identifying the target species that should benefit. Based on the spatial
scale of Europe, and knowledge regarding migration, dispersal capacity and
behaviour, an assessment was made of which species might be able to benefit
from 'European' corridors.

This study concentrates on vertebrate species and not without reason. A
functional approach with an attempt to generalise and classify, as presented
here, can at present not be formulated for plants and invertebrates without
either going into detail for the many ecologically different species groups or
grossly simplifying the matter and thus losing scientific accuracy. We cannot
do justice to the enormous variety within these species groups within the scope
of this chapter. Particularly in the case of invertebrates, studies concerning corridor function are scarce (Speight 2000; Vermeulen 1995).

In this chapter, therefore, a species-specific typology of corridors for vertebrates is presented, as well as an analysis of the potential target species of
European corridors for (endangered) vertebrates.

6.2   Corridor terminology: a species-specific approach

The terminology on corridors is highly variable and inconsistent: habitat corridor, dispersal corridor, movement corridor, wildlife corridor, biocorridor, landscape linkages and greenways are being used in various contexts
(Buček *et al.* 1996; Saunders and Hobbs 1991; Simberloff *et al.* 1992). Every
country or region also has its own set of definitions (see also Box 2.2).

Reviewing the literature on connectivity, it appears that the term 'corridor'
is used in reference to two different processes occurring in the landscape:

- biotic processes: landscape structures that facilitate or direct the movements of animals or seeds of plants and thus improve connectivity (Bennett 1998; Jongman and Troumbis 1995)
- abiotic processes: landscape structures that facilitate, direct or influence the movement of fluxes of water, energy and materials (Bridgewater 1987; Kavaliauskas 1996)

In this chapter, we restrict ourselves to a typology based on biotic processes. Most scientific research on the topic of corridors has been focused on corridors as a means to enhance connectivity for species.

The definition of a 'corridor' as a structure that increases movement or connectivity of species varies. Saunders and Hobbs (1991) define a corridor as a *linear* feature of vegetation, which differs from the surrounding vegetation and connects at least two different patches that were connected in historical times. Research into dispersal mechanisms for forest birds showed that certain species do not rely on linear features for their dispersal but do need a mosaic pattern ('stepping stones') for dispersal (Brooker *et al.* 1999). Bennett (1998) distinguishes three different types of corridor based on physiognomic structure: linear, stepping stone and landscape. A linear type of corridor consists of a continuous or near-continuous link of suitable habitat; a stepping stone corridor consists of suitable habitat interspersed with disturbed or unsuitable habitat. Landscape corridors consist of a range of more or less different suitable habitats in which mosaics occur instead of rather sharp discontinuities (see Figure 6.1). In line with the definition given by Vos *et al.* (2002) and Bennett (1998), corridors are defined here as:

> A functional linkage between resource habitats of a species, consisting of landscape structures that are different from the matrix resulting in a favourable effect on the exchange of propagules (individuals, seeds, genes) of the species.

As the main function of a corridor is to enhance or maintain the rate of exchange of propagules, corridors are species-specific and can have a variety of functions (Beier and Noss 1998). Hence, it is only possible to distinguish a corridor if exchange occurs along specific routes and if movements are bound to specific landscape features. For species that do not show specific preferences for certain landscape structures or habitat types during dispersal, migration or daily movements, no corridors can be distinguished.

It is important to note that we see a corridor as a function rather than as a structure. Therefore function was considered to be the most important feature for setting up a corridor typology. Based on the function of a corridor, three different types have been distinguished: migration corridor, commuting corridor

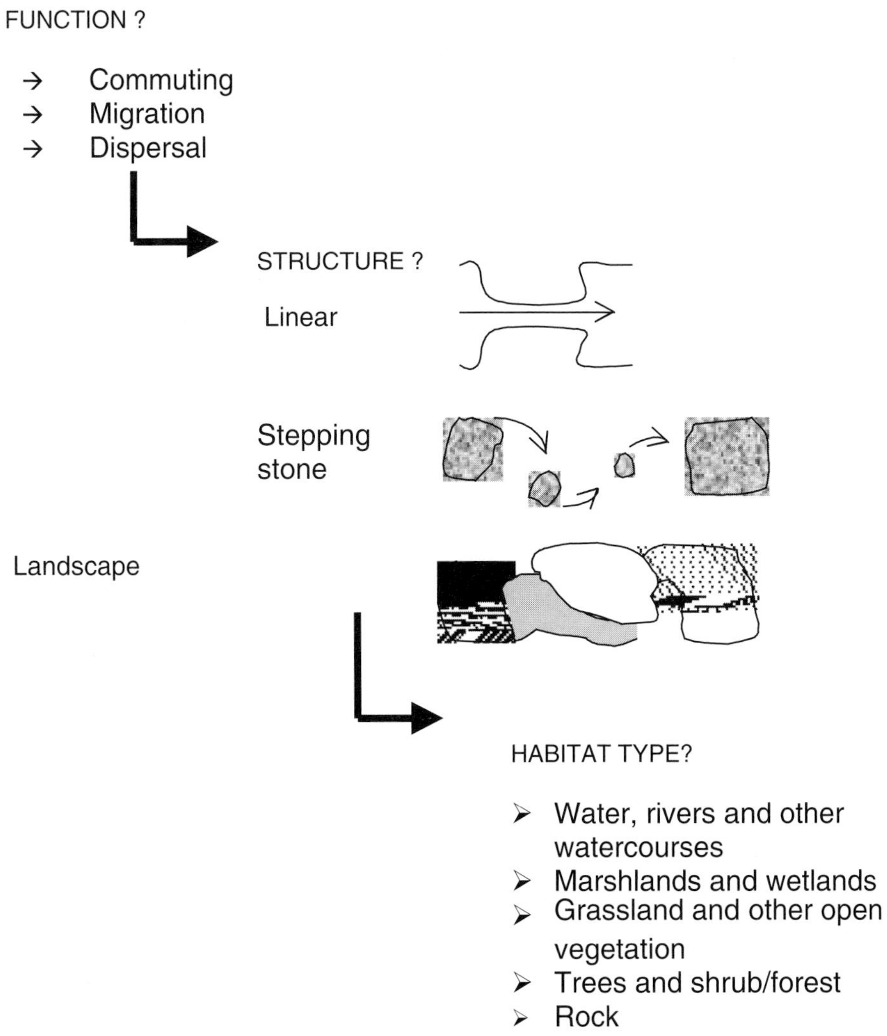

FUNCTION ?

→   Commuting
→   Migration
→   Dispersal

STRUCTURE ?

Linear

Stepping
stone

Landscape

HABITAT TYPE?

➢   Water, rivers and other
    watercourses
➢   Marshlands and wetlands
➢   Grassland and other open
    vegetation
➢   Trees and shrub/forest
➢   Rock

FIGURE 6.1.
Typology of corridors based on function (migration, commuting and dispersal) and
on physiognomic features and habitat types.

and dispersal corridor (Opstal 1999). A further division is made based on physiognomic features and habitat types (Figure 6.1).

The three types of corridor that have been distinguished are as follows.

1. **Commuting corridor**, used for daily movements from resting or breeding
   site to foraging area. A commuting corridor links elements, within a home
   range of a species, that have a different function. It supports daily movements between these elements and acts beneficially because, for example, it
   reduces predation risk, offers guidance and facilitates movement through

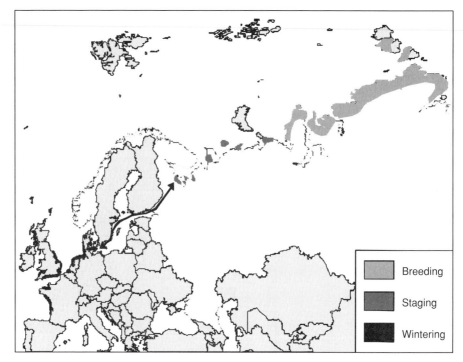

FIGURE 6.2.
Example of migration corridor: brent goose migration and stopover pathway. This map shows that this migratory bird species uses stopover (staging) sites in a quite narrow band between its breeding and wintering range (courtesy of Bart Ebbinga, based on Green *et al.* (2000, 2002).

the landscape. Normally these movements are restricted to short distances (up to a few kilometres) for vertebrates, or to tens of kilometres for wider-ranging species. Good examples of species using commuting corridors are badgers and bats (Limpens and Kapteijn 1991; Broekhuizen 1986).

2. **Migration corridor**, used for annual migratory movements from one resource area to another (e.g. from breeding to wintering site). The biological process of migration is very prominent in many species groups. Best known are bird and fish migrations. In their journey from one resource area to another, some species will benefit from the use of corridors. This can be in the shape of a continuous linear pathway (as for riparian fish species). More often the pathway will consist of a set of areas used during migration as 'stopover' places, for example marshes for waterfowl and waders. An example of a migration corridor for waterfowl is the migration route of the dark-bellied brent goose (*Branta bernicla*) (Figure 6.2).

3. **Dispersal corridor**, used for a one-way movement of an individual (usually a juvenile) or population from either its former breeding area or its site of birth (for juveniles) to a new breeding area. Dispersal is an essential

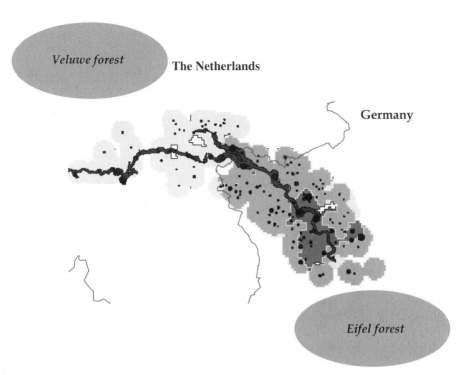

FIGURE 6.3.
Reproduction corridor for the middle spotted woodpecker (*Dendrocopus medius*) in the Rhine Valley. The map shows the floodplains of the river Rhine crossing the border of Germany and the Netherlands. The black dots indicate the location of habitat sites (ancient oak forest). The grey shades represent the probability of occurrence in a region. The darker the colour, the higher the probability.

process leading to the immigration of individuals into other populations or to (re)colonisation of suitable habitat. Populations in fragmented habitats can function as a metapopulation: as long as the fragmented habitat patches are not totally isolated from each other, they constitute one metapopulation in a habitat network (Hanski 1994; Opdam 1990). In order to differentiate between individuals and populations, dispersal corridors are subdivided into three types: one-step dispersal corridors, reproduction corridors (Figure 6.3), and range expansion corridors.

One-step dispersal corridor
The corridor serves to facilitate the exchange between two habitat patches that are located within the dispersal range of an individual. The aim of this corridor is to increase cohesion of patches *within* one habitat network. The quality of the corridor should allow individuals to disperse but reproduction in the corridor is not necessary. As soon as the distance between suitable habitat patches

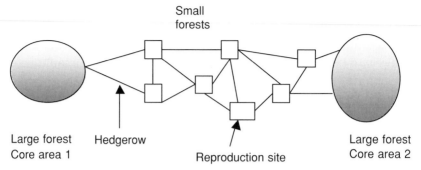

FIGURE 6.4.
Theoretical example of a combined reproduction and dispersal corridor for a ground-dwelling species.

becomes too large there is a risk that the corridor will act as a sink; dispersing organisms will not be able to reach an adjacent suitable habitat patch.

Reproduction corridor

The reproduction corridor establishes a link between two habitat networks. It consists of a system of habitat patches to fill the gap between two (formerly unconnected) habitat networks. Since the distance cannot be bridged by individuals in one leap, for instance by an individual within one year, a system of reproduction sites is needed that supports the exchange of individuals. The quality of the habitat in the corridor should enable reproduction of individuals in the corridor. An example of such a reproduction corridor for the middle spotted woodpecker is represented in Figure 6.3.

Range expansion corridors

As a result of climate change it is expected that the distribution range of many species will change. Corridors have been suggested as a means to alleviate the adverse effects of climate change on species. Corridors would provide routes for species to be able to shift their distribution as a response to climate change.

In some cases a species requires a combination of a dispersal corridor and reproduction corridor. In Figure 6.4 a theoretical example is given for a ground-dwelling forest species.

6.3   Need and potential of corridors with a European dimension

To assess the need for corridors to increase connectivity for species on the European scale, the spatial and temporal scale under consideration guides the analysis. Data on postglacial colonisation processes of immobile species

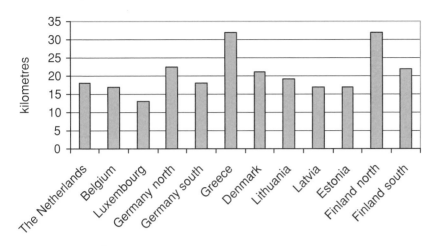

FIGURE 6.5.
Mean distance in kilometres between CORINE Biotope sites in ten European countries. Germany and Finland have been divided into northern and southern parts. Several sites (10–20 per country) were selected at random to calculate the mean distance. For each selected site the distances to the nearest five CORINE Biotope sites were calculated and averaged. The figure given for each country is the average distance of all selected sites within that country. The mean interpatch distance for these sites is between 13 and 32 kilometres. This calculation did not take into account the type of habitat occurring within the CORINE site or the size of the site. Therefore it is most likely that the actual distance between similar habitat types will be larger then the figures given.

show that these processes take several hundreds of years. Although in the long run immobile species could benefit from European corridors, in the short and medium term, only species with a large or medium dispersal capacity will be able to benefit from the new opportunities offered. Practically, we have selected species that show movements of at least tens of kilometres.

A preliminary analysis showed that this distance coincides with the interpatch distances between important European nature sites (Figure 6.5).

Corridors are promoted as a tool to improve the conservation of species. Although corridors on a European dimension might also prove to be beneficial for a broader range of non-endangered species, present conservation efforts are aimed at endangered vertebrate species. Several conventions, agreements and articles of EU legislation were reviewed to be able to select which European endangered vertebrates could benefit from corridors with a European dimension. All vertebrate species listed in Annex 2 of the Habitats and Species Directive, or Annex 1 of the Birds Directive, and vertebrate species mentioned as target species for the Emerald Network under the Bern Convention and Appendix 1 of the Bonn Convention, were considered in this analysis.

In total 420 vertebrate species (including some subspecies) were reviewed. Species were selected on the basis of the criteria described in Table 6.1. The

Table 6.1. *Criteria for selection of species that might benefit from European corridors*

Criterion 1: functional use

1. In the case of migrating species only species using specific flyways or routes were selected
2. In the case of commuting and dispersal movements only species that need specific landscapes structures to facilitate movement were selected

Criterion 2: movement ranges

1. Species with a dispersal range exceeding 30 km were selected
2. Species with a dispersal range exceeding 10 km and a small minimum area requirement for reproduction (<10 ha) were selected as species benefiting from reproduction corridors

Table 6.2. *Total number of vertebrate species mentioned in the lists and the percentage of the total number of species (per taxonomic group) that benefit from corridors on a European scale*

Species group	Total number of species in lists	Percentage of species with sufficient movement range	Percentage of species that could benefit from 'European' corridors[a]
Mammals	59	83	39
Birds	204	93	34
Herpetofauna	58	0	0
Fish	99	12	12

[a] Excluding marine vertebrates and marine herpetofauna.

outcome of the analysis is presented in Table 6.2. It is estimated that, of the species reviewed, 104 could benefit from European corridors. Of these 104 species 69 are birds, 23 are mammals (mostly large herbivores and carnivores) and 12 are fish. Mammals require most of all dispersal corridors, fish need migration corridors, and birds require both dispersal and migration corridors.

Information on the distribution of all selected species was gathered based on the biogeographical regions map of Natura 2000 and its extension to Central and Eastern Europe. An overview of the distribution of the identified species in the biogeographical regions is given in Table 6.3. The highest number of potential target species occurred in the Mediterranean and Continental region.

A further analysis was carried out in order to gain insight into the habitat required by the target species in the corridor. Table 6.4 presents an overview of the habitat types preferred by the selected species.

Table 6.3. *Distribution of the selected species according to the biogeographical regions classification*

The figures are percentages of the total number of species. The most striking features are presented in bold.

	Number	Sea	Arctic	Mediterranean	Macronesian	Alpine
mammals	23	0	24	**36**	0	28
birds	69	9	16	**51**	7	28
fish	12	50	17	42	17	17
TOTAL	104	11	18	**46**	7	26

	Atlantic	Continental	Boreal	Pontic	Anatolic	Pannonic	Steppic
mammals	16	28	**32**	12	16	8	8
birds	29	41	35	20	7	32	36
fish	**50**	**58**	42	17	8	42	33
TOTAL	28	**39**	35	18	9	27	29

Table 6.4. *Preferred habitat type of the corridors for the selected species*

The figures are percentages of the total number of species. The most striking features are presented in bold.

	Number	Watercourses	Marshlands	Grasslands/ open terrain	Shrub/ forest	Rocks/ mountain	Open sea
mammals	23	16	8	**44**	**44**	12	0
birds	69	4	6	28	**41**	9	6
fish	12	**100**	0	0	0	0	0
TOTAL	104	18	6	28	**36**	8	5

The analysis of the need for corridors based on species' requirements shows that many vertebrate species could benefit from European corridors. Mammals benefit from permeable landscapes built up by mosaics of semi-natural and natural ecosystems, functioning as a dispersal corridor. Birds benefit from both migration and dispersal corridors, but are probably less bound to the physical features. Fish obviously benefit from linear corridors such as large rivers.

According to this study mosaic landscapes consisting of forests and extensively used grasslands have a high potential as dispersal corridors for many vertebrate species provided that no large barriers (infrastructure, urbanisation) occur. The land use of these areas can be multi-functional if biodiversity considerations are taken into account.

6.4   Discussion

The typology and methodology presented in this paper offers a tool to
assess the need for corridors on a European scale. However, the typology and
methodology has its limitations in so far as it only applies to terrestrial and
freshwater vertebrates. For plants and invertebrates a different or adapted typology is needed. Plants and invertebrate species should probably be classified
more according to their dispersal *mechanism* than by the function of movement.
The mechanisms involved in dispersing individuals or propagules from one
site to another will determine whether or not species can benefit from corridors. In plants the dispersal mechanisms do not necessarily reflect the actual
dispersal. For these species groups sudden and periodic events and features also
should be taken into account.

For wind-dispersers it is hard to imagine any functional use of designed and
managed corridors. However, a fair number of invertebrate species have dispersal mechanisms comparable to those of vertebrates, for example butterflies and
carabid beetles. For these species some elements of the chosen approach might
be useful.

The indicative analysis of target species for European corridors presented in
this paper is a rather simplified analysis for assessing which species according
to their dispersal capacity and migration patterns are able to move over distances exceeding 10–30 km. Although this is a quite arbitrary figure, it is in line
with the analysis on the interpatch distances of CORINE Biotopes. It is feasible
that these species will use migration, commuting and dispersal corridors that
have a 'European' dimension (i.e. crossing national borders) and can therefore
be considered as target species for European corridors.

The question of whether these species would indeed benefit from the safe-
guarding or restoring of European corridors can only be answered after a more
detailed analysis. Such an analysis should take into account the current distribution range and the size and configuration of the current habitat.

Another reason for the ongoing scientific debate has been the tendency
to discuss the need for corridors without a proper problem analysis and
discussion of alternative solutions. Developing corridors is not an *aim* as
such, but merely one of the tools available to conserve species. However,
owing to the popularity of corridors, sometimes the opposite is true. Some
corridor plans have been justly criticised for the limited problem analysis that underpins them. Simberloff *et al.* (1992) give several examples of
huge investments in corridor projects for which the aims are not clearly
specified.

Habitat distribution, population viability and species habitat and corridor
requirements define whether a corridor in a certain situation is necessary or

not, and will contribute to the conservation of a specific species or species group (for discussion on corridors, see also Beier and Noss 1998; Bennett 1998). In judging whether or not a corridor is an efficient measure to undo the detrimental effects of fragmentation, one should take into account not only the potential beneficial effect of corridors, but also the effectiveness of other measures that will increase the conservation status of a given species. For some species other measures, such as restoration of habitat in new or existing nature areas, proper management of existing sites or translocation of animals in combination with captive breeding and reintroductions, might be a better option to ensure their conservation. Often a combination of measures is needed to ensure conservation. Financial considerations also need to be taken into account to ensure that the proposed corridor is the most cost-effective solution in respect to other ways of using scarce conservation resources.

In many (greenway) projects corridors also have other functions for society (amenity, hiking, water transport). This means that the aims and benefits of the corridors should be clearly defined.

Within the European Union the Natura 2000 network is currently being developed. This network is one of the most important implementation instruments for the development of the Pan-European Ecological Network. Until now limited attention has been paid to the spatial configuration of this network. We plead for a consistent evaluation of the Natura 2000 network in order to assess whether the network will be sustainable in the long term. This analysis should include an assessment of the value and effectiveness of corridors on a European scale.

6.5   Conclusions

Recent scientific studies have underpinned the intuitive feeling of the beneficial use of corridors for vertebrates (Beier and Noss 1998). However, for corridors to actually function, one should keep in mind the ecological needs of the species. It is essential to perform species-specific problem analysis that takes into account the configuration of the entire habitat networks in order to define and locate corridors.

The typology developed in this paper offers a way to describe connectivity for vertebrate species on a European level. For plants and invertebrates, different approaches need to be developed. The analysis presented shows that many vertebrate species, mostly with a large dispersal capacity, could benefit from 'European' corridors.

According to this study, mosaic landscapes consisting of forests and extensively used grasslands have a high potential as dispersal corridors for many vertebrate species provided that no large barriers (infrastructure, urbanisation) occur. The land use of these areas can be multi-functional if biodiversity considerations are taken into account. With the typology and results of the analysis presented in this chapter, a method is offered to identify corridors in the Pan-European Ecological Network.

7

Planning the future landscape between nature and culture

7.1 Introduction

The concept of ecological networks stands within the discourse of nature conservation and land use planning. The rapid changes brought by human activities to the landscape have caused fragmentation of the natural environment and isolation of species and habitats. The degradation and even destruction of these natural habitats are nowadays a common issue (see Bennett 1999) and open up the debate on biodiversity and landscape conservation, natural resources consumption and sustainable development.

Recent studies, however, have indicated how ecological networks in Europe are related to biogeography, population dynamics, landscape ecology and land use science (Jongman 1995, 1998a; Pungetti 1999). There is indeed strong evidence that they do not consist only of ecological elements, but also of cultural components. Nature and culture become thus a key duality in the approach to nature conservation and sustainable development for the new millennium.

In Italy nature and culture retain their own particular value and weight, are both fundamental in environmental sciences, but are often in conflict. In past traditional practices there was a balance between nature and culture (Pungetti 1995, 1996), but when land use activity declined nature expanded, with several consequences. Ecosystem re-establishment, for instance, often took place with a reduction in biological and landscape diversity, and although there was ecological improvement in rural areas, the cultural landscape vanished.

For the development of ecological networks in areas of cultural landscape (see Kubes 1996) the above aspects become imperative (Figure 7.1). In these areas the link between natural habitats and human practices is quite evident. The continuous interaction between nature and culture in Italy is the reason why this country has been chosen for the study of ecological networks within this context.

FIGURE 7.1.
Sagittario Valley in the Abruzzo National Park: a typical example of the territory in the Central Apennines, with large natural areas, important cultural features and infrastructural elements.

This chapter explores ways of planning ecological networks in areas where the cultural landscape shows strong links between natural and cultural elements. Starting from considerations of environmental planning in Italy, it analyses the national biopermeability and environmental continuity and provides an overview of the development of the ecological network concept in the country. It proposes a preliminary recognition of the biopermeability conditions in Italy, focusing on three categories: hydromorphology; agriculture and forestry; barriers to biopermeability. It also suggests a shift of focus from the design of the national environmental continuity to the development of ecotones, studying the relationship between environmental continuity structure and the single-species-oriented ecological network. It finally advances the concept of 'de-anthropisation of the plan' and urges the integration of ecological networks into the planning system, employing both ecological and social considerations.

7.2   Nature and culture

Cultural heritage is diffused all over Italy in forms that are different from, but at the same time interconnected to, the environmental structure. Being an element of landscape enrichment and quality, cultural heritage offers on the one hand an added value to the territory. This is particularly true when cultural heritage shows the historical value of a certain human development and in this way enables the conservation of nature. On the other hand, the cultural heritage may constitute a barrier to the environmental structure, for example when it leads to intensive tourism disturbing the ecological context. Clearly, almost the whole Italian landscape, like a great part of the European landscape, has been shaped and deeply marked by people. These instances suggest that in Europe it is necessary to take into account the cultural heritage in both environmental policy and ecological network development.

Accordingly, future land use and land disuse cannot avoid considering landscape evolution. Historical analysis, culturally and ecologically oriented, thus becomes an essential element for decisions on either future development or re-naturalisation (Makhzoumi and Pungetti 1999). Until 50 years ago landscape changes in Italy, though directed by human interventions, did not upset the environment. Today landscape changes can bring about major disruptions. They, moreover, are not only artificial but can also be spontaneous. It is not certain, however, whether they can positively influence the natural environment.

From here, a scientific debate arises; it relates to the uncultivated agricultural areas that once made up the Italian cultural landscape and are now abandoned. Having lost their productive importance and not retaining a natural value, these areas need a new identity and collocation. Bringing them back to their traditional use is no longer sustainable, unless policies are applied addressed to territorial sectors of peculiar interest. Letting them develop, via spontaneous transformation, into a natural environment is an interesting prospect, though uncertain to succeed.

A similar situation occurs with cultivated upland areas, which in Italy are often on terraces with poor soil, fruit trees and shrubs. Spontaneous reforestation is here recurrent but not always pleasant, especially in the early stage with no human control. This type of re-naturalised cultural landscape, nevertheless, might be important for the ecological restoration of the Italian landscape. However, scientific evidence of this importance is still lacking and further studies of such areas are necessary to prove their potential ecological value.

The Italian cultural heritage furthermore includes, apart from the cultural landscape, other elements of historical importance and visual value, such as historical centres, isolated monuments and ancient tracks. Though these do not interfere with an ecological network, they are present in the territory and could become a barrier if misused. A further consideration of nature and culture will be provided in the discussion section.

7.3   The de-anthropisation of the plan

Environmental planning in Italy has often considered natural components separately from cultural ones, thus ignoring their interaction. In addition, it still emphasises conservation of natural habitat rather than nature development. This trend has confined environmental planning in Italy to territory and parks, without addressing it to the wider countryside and ecological systems.

In such a context it has been necessary to review the planning theory with a new concept that we define as 'the de-anthropisation of the plan'. It consists in giving less weight to the anthropogenic features and more to the ecological characters of the site. Moreover, it considers the plan as a useful tool to know and improve the conditions not just of the human, but also of the abiotic and biotic components of the landscape.

It is known that habitat fragmentation constitutes one of the first causes of biodiversity decline. It is therefore urgent in Italy to outline the national and local characters of environmental continuity that can create the foundations for planning activity. The latter should aim at solving those conflicts between the human use of the territory and the environmental continuity that lead to environmental fragmentation. Two phases are proposed to tackle the problem. In a first phase, environmental sciences can be linked to physical planning in order to analyse the territory globally. The abiotic complex, the biotic complex and the socio-economic complex can be studied in relation to the occlusion or disturbance they cause to the land. In a second phase, a specific design is required for interventions on defragmentation. This can be carried out using a large variety of disciplines, from both geographical and natural sciences.

7.4   Biopermeability and environmental continuity

Biopermeability regards the complex geography of the natural territory. In a cultural landscape the study of biopermeability has to be addressed considering the link between natural and cultural areas. The latter retain species able to adapt to the human environment, while the former retain a greater variety of species, including those unable to adapt, that is, the most threatened ones.

Biopermeability spaces are non-urbanised parts of the territory or areas with low urban density including intensive agricultural areas. The big urbanisations (e.g. highways, linear settlements, industrial areas) form barriers to potential species movements. Overlaying of urbanised areas with the geography of biopermeability allows assessment of the territorial units of 'environmental continuity'. These are in turn the ambits where it is possible to implement efficient defragmentation interventions within acceptable margins of economic and technical effort (Romano 2000).

In the first instance the environmental continuity structure, derived from the biopermeability areas, rests on urban components and land use activities.

In the second, the ecological network rests on biological and natural components. Biopermeability and environmental continuity structures, however, do not always coincide with the ecological network structure.

The analysis of the national environmental structure in Italy, moreover, brings evidence of the aforementioned lack of interest in natural aspects by the past planning system. Having been often confined to administrative borders and to national strategies, the planning system has been unable to conceive the geographic, morphologic and environmental continuity of large ecosystem units such as the Alps and the Apennines.

The infrastructure managers in Italy, furthermore, have long disregarded environmental concerns. Only recently, under the pressure of the European Community, have they started to consider mitigation aspects of human impacts (e.g. pollution, noise, urbanisation) and to apply them to a few isolated cases of environmental defragmentation, e.g. the Tarvisio Forest.

It is nevertheless true that since the 1970s national instruments such as the Mountain Map (Ministero dell'Agricoltura e Foreste 1976) have allowed us to visualise the levels of geographical continuity in the natural and semi-natural areas of Italian uplands, and the effects that infrastructures and urbanisation caused to the environment. The reading of such instruments, utilising GIS, enabled us to outline the situation as an almost continuous configuration of areas with at least a minimum level of biopermeability, overlaid by the infrastructure network that interrupts the Apennine continuity in a dozen macro-zones called Efficient Units of Environmental Continuity, or EUEC (Figure 7.2). Within these it is still possible to obtain efficient eco-connections if ecological conditions are supported by sound economic and technical efforts.

It is generally recognised that protected areas that do not fit within the environmental continuity structure do not offer positive results in the long term. This is even more evident in a country like Italy, where the extent of the largest park reaches 200 000 hectares and is therefore too small for the medium home range of many wildlife species. Another paradox is the presence in the country of endangered species such as the brown bear (home range of 70–100 km^2) (Boscagli 1988), the Apennine wolf (70–200 km^2) (Boscagli 1985) and the royal eagle (100–160 km^2 for a pair) (Spinetti 1997). In this context, strategies for environmental continuity have to expand to the wider landscape, where the presence of marginal areas, such as degraded and uncultivated land, assumes strategic importance.

The core system of protected areas, in addition, has to include the peripheral system, i.e. the ecological network. On this line, research on the planning of ecological networks has demonstrated how the protected areas have different characteristics in relation to the national environmental continuity

100   0   100   200   300 kilometers

NATIONAL PARKS

OTHER PROTECTED AREAS AND SITES OF EUROPEAN INTEREST

LARGE NATIONAL UNITS OF ENVIRONMENTAL CONTINUITY

FIGURE 7.2.
National map of environmental continuity of Italy.

structure (Romano 1999). As a matter of fact, territorial changes have brought about alterations to the national geographic condition that can be easily recognised on the territorial maps.

Furthermore, protected areas in Italy do not always present environmental continuity within their territory. Some (e.g. Pollino Park) are internally fragmented because of infrastructure or production activities, while others (e.g. Cilento–Vallo di Diano and Gargano Parks) cannot be linked with the surroundings because of large barriers. In a few cases it is even impossible to connect two very close parks (e.g. Monti Sibillini and Gran Sasso – Monti della Laga Parks), owing to adverse morphology, intruding infrastructure and industrial areas.

The study of the location of marginal natural areas, instead, has shown other directions of minimum resistance to biological movements (Boitani 1997). These directions coincide with the contiguity lines, useful in park planning activities (Shafer 1999). From the above it is evident that the Apennine parks play a fundamental role to assure, via a sound environmental policy, the biopermeability and the environmental continuity of the Italian Apennines.

7.5   Planning ecological networks

The ecological network concept in Italy has only recently been introduced and therefore specific criteria have not yet been developed. At the national and regional level, for example, there is a lack of legislation on ecological networks, but related matters can be found in acts and directives on nature conservation, as in other European countries. At subregional level, by contrast, a few provinces with the support of academic scientists have produced ecological network studies.

The Italian planning legislation does not yet consider environmental connections. Only a few regional urban planning laws (e.g. in Basilicata and Emilia Romagna) refer to ecological networks and corridors. A new approach was initiated by the National Agency for Environment Protection (ANPA) by setting up the Ecological Network Action Plan. It aimed firstly to define instruments for ecological continuity and methods for ecological network development, and secondly to integrate ecological networks into physical planning. For its implementation a national working group was set up and ten case studies were chosen, carried out by local working groups consisting of academics, regional authority scientists and local authority officers.

Following the development of ecological network initiatives and methodologies of other European countries, in the past few years Italian scientists too have tackled the topic, but have recently faced two main problems related to the aspects already illustrated. First, there is the complex interaction between natural and cultural systems (Figure 7.3), owing to the strong historical and human presence, and secondly there is the neglect of environmental and ecological components in the legislative and planning systems. As a consequence environmental conservation and development have been pursued in two disconnected areas, i.e. nature protection and physical planning, that unfortunately have not yet been integrated with each other.

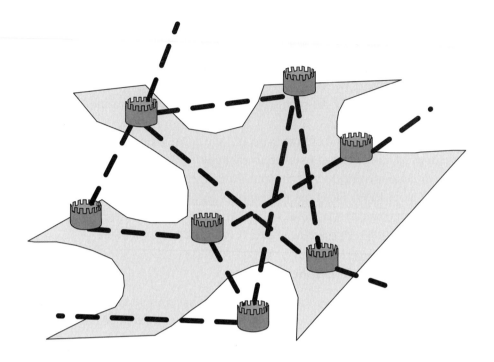

Biopermeability areas (environmental continuity system)

Historical and cultural elements

Historical trails and other cultural connections

FIGURE 7.3.
Relationship scheme between the environmental continuity system and the cultural system in Central Italy.

One of the most significant scientific pioneer approaches tackling the subject of planning ecological networks is PLANECO (Planning in Ecological Networks). PLANECO is a research project founded by the Italian Ministry of University and Scientific Research for the years 1998–2000. It involves five universities: L'Aquila, Camerino, Chieti, Roma 3 and Molise. Among its goals is to define planning criteria for protected areas that are not isolated but linked as components of an ecological network as well as for the territory considered in a coherent ecological system with particular attention towards environmental continuity.

In Italy the interconnection between the natural and cultural systems, together with the overlay of a national ecological network, interferes at all the levels of planning and land use transformation. The PLANECO Project aims at contributing to the solution of this problem in three ways: (1) through a discussion on the role of the plan, useful to individuate, conserve and restore environmental continuity; (2) by developing directions to orient land use within protected areas, considered as core areas of an ecological network; (3) by proposing new possible urban development forms, allowing connective functionality of the suburban areas (Romano 1999).

In the first case, 'dynamic indicators' will be used to determine the present situation and possible scenarios. Among these indicators are settlement dispersion, infrastructure density and permeability, and spatial and quantitative relations between natural and urban land use (Di Ludovico and Romano 2000). In the second case, the new concept of a ramified zone structure of protected areas takes the place of the old concept of a centralised zone structure. This allows a better environmental continuity between the inside and the outside of the parks, and supports an ecocentric planning process, to be tested in areas of special environmental protection. In the third case, town planning principles such as linear urban development are reviewed because they fragment environmental continuity and threaten the quality of nature.

One of the first results of the PLANECO Project is the identification of the elements connected to environmental continuity at national scale. The CORINE Land Cover data have been used to elaborate the National Biopermeability Map. This evidences the geography of land use to which the different levels of nature and the most relevant national environmental systems can be linked. An uncertain role is played by the agricultural areas, but further data are necessary to elaborate their specific 'fragmentation indicators' (e.g. related to geometry, texture and conductivity). The analysis of the PLANECO research at the moment is therefore based mainly on environmental systems and land use.

7.6   Assessing environmental continuity and fragmentation

In the preliminary recognition of biopermeability conditions in Italy, research analysis focuses on three major categories based on land use features: (a) hydromorphology; (b) agriculture and forestry; (c) barriers. Macro-elements are indicated for each of them, determining not only the characters of the biopermeability for a whole area, but also additional aspects, which can be studied in detail in order to individuate the eco-connectivity lines.

There is, however, a difficulty in proposing a hierarchy of biopermeability for homogeneous land use areas. For biodiversity assessment it is indeed fundamental to analyse the configuration and distribution of ecotones.

7.6.1 Hydromorphology

This major category is characterised by landforms and hydrological features and has been divided into two macro-elements: morphology and hydrology.

In environmental evaluation regarding biological movements, morphology is essential to verify potential environmental continuity to which the presence of natural barriers or linkages has to be referred. Specifically, a study of the land structure and soils can be carried out to establish the presence of landforms favourable to connections.

Clear morphological lines, i.e. without many transverse obstacles, facilitate the movement of species (Figure 7.4). On the other hand, a more articulated type of morphology facilitates a very high level of biodiversity owing to a variety in microclimate, cliffs and slope exposure. The morphological elements, nevertheless, are not suitable for a generic assessment of carrying capacity at a large scale. They require in fact an assessment at a smaller scale, which allows a better consideration of ecological specifications and details.

Among the hydrological elements, rivers are the most difficult to assess for carrying capacity at general level. The rivers are certainly the optimal linkage for the movement of species living both in the water and on riparian land (Jongman 1998b). In the North–Central Apennines the river conditions have radically changed, however, and high environmental quality can be found only in the uplands. But even here, in agricultural and urban areas, the quality is lower, especially in the water, and many rivers have to deal with human impacts from river works and quarry extractions.

At present Italian rivers can function as ecological corridors only in those upper parts of their course where the environmental quality reaches a reasonable level. Close to the coast, on the other hand, the rivers do not retain any connective function owing to the numerous human barriers. With the high concentration of pollution, canalisation, urbanisation and delta degradation, rivers are actually barriers themselves. Therefore the hydrological elements, like the morphological ones, require separate studies and small-scale environmental assessment in order to define the possibility of being active parts of an ecological network.

7.6.2 Agriculture and forestry

This major category is characterised by land use types and has been divided into four macro-elements: agricultural areas, uncultivated and degraded areas, forest areas and pastoral lands.

Italian agricultural areas are generally ecologically unfriendly because intensive farming, farmhouses, noise and illumination impoverish biodiversity. To add to this there are the barriers created by property fences, human movements and infrastructural networks (Jaarsma 1997). The level of biopermeability can be very low and consequently many of these areas are not suitable to function as ecological corridors. Despite this, a few uplands with extensive agriculture

MORPHOLOGICAL STEPS

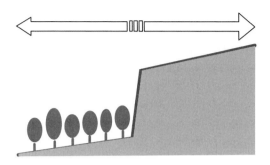

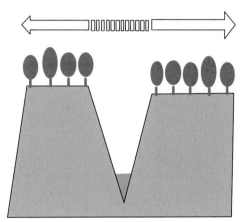

CANYONS

RIDGE STRUCTURES

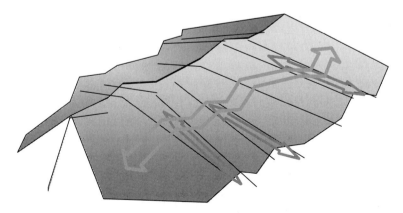

FIGURE 7.4.
Examples of environmental continuity according to different morphological situations. The broken parts of the arrow indicate areas of reduced continuity.

present a higher level of biodiversity and could already be used as ecological corridors.

Studies on these areas have shown the difficulty of assigning to them a proper level of biopermeability unless a detailed study is carried out. In large areas of agricultural land the connectivity is poor, but depending on the type of cultivation, field dimension and productive cycles, variation between areas is considerable.

However, agricultural areas that are nowadays ecologically unfriendly can have the potential to become a resource for species movements, changing their barrier function into connection. Their landscape structure is relevant in this aspect. Long and narrow configurations, for example, present transversally high biopermeability and facilitate the biological movement between the near areas with a low degree of disturbance.

Uncultivated areas, with residual agricultural vegetation, present on the average a good level of biopermeability since they do not retain either barriers to species movement or disturbance by human impact. The debate on marginal lands in Italy has been based on the equation that lack of explicit and localised environmental quality is equal to the constraining environmental impact of the planning actions. Yet scientists have proposed for these areas the alternatives of either non-intervention or urban development.

Marginal lands have had little consideration in the Italian framework of environmental quality. These areas, however, retain a particular importance regarding links with the surroundings and can hence form a valuable element in an ecological network. Moreover, owing to their low economic value they are priority sites for environmental restoration and re-naturalisation.

The level of biodiversity in uncultivated and degraded areas varies from site to site and necessitates a detailed examination of their physical, ecological and structural characteristics. The aspects that can be considered in biodiversity assessment are the typology of previous land use, period of abandonment, vegetation, altitude, slope exposition, geometric forms of relief and specific local factors.

In Italy forest areas present high ecological values for many animal species, owing to the spread of refuges and hiding places and to the generally limited human disturbance. A high level of biodiversity can be found in this macro-element. More detailed research, however, is needed to identify differences in regards to parameters and indicators for biodiversity and biopermeability (e.g. forest texture, species diversity, past and present levels of human use, local morphological structure, disturbance factors). This further evaluation should present a wide spectrum of biopermeability levels, fundamental for the identification of environmental continuity.

In Italy pastoral land is a macro-element with an acceptable level of biodiversity to enhance environmental continuity. The vegetation diversity here is often higher than in uncultivated or degraded lands, although in pastoral land the disturbance due to human activities is higher. If blocking of environmental continuity, e.g. through property boundaries, occurs in these areas, it can mostly be easily overcome owing to its incomplete and non-prominent character and low impact on the territory.

7.6.3 Barriers to biopermeability

This macro-category is divided into three macro-elements: the urban system, large complex barriers, and simple barriers. Planning biopermeability at a large scale and design of specific projects clearly influences infrastructure policies and project realisation.

Urban areas are barriers that cause a total blocking of ecological connectivity. Here artefacts, disturbances and physical obstacles to biological movement are dominant. The geographical distribution of settlements is fundamental in the mapping of environmental fragmentation. A linear settlement has higher impact on fragmentation than a concentric settlement. Even with similar area, the first type of settlement has longer infrastructure elements and a wider spread of illumination, noise and movements than the second type.

Possible environmental continuity lines have to be found in those urban elements that present a minimum level of naturalness and connectivity, e.g. continuous green areas, rivers and riversides. Although comparable to the American greenways (Little 1990) these areas in Italy contain a very limited degree of naturalness and a high human impact that has to allow linkages between the urban areas and the natural and semi-natural areas outside the city. In general the role of these areas is limited to recreation and to the conservation of those species adapted to urban environments.

There are, however, examples of settlements in the uplands, e.g. L'Aquila and Sulmona in Abruzzo, which are well integrated into their natural and semi-natural environments. Hence here new forms of urban planning and management that include connectivity can be experimented on. Opposed to this, in the same region, is the complex urban concentration along the Adriatic coast that together with the connected system of infrastructure (Figure 7.5) presents an insurmountable obstacle to the ecological continuity between mountains, river valleys and sea. The natural recovery of these areas appears unlikely since the political, technological and financial efforts that environmental continuity requires for coastal areas are enormous. The solution at the moment is to conserve those rare non-degraded situations and preserve them from the human impact described above.

The characteristics of an urban system in relation to environmental continuity can be pointed out through the use of sound indicators of the link between urban areas and the countryside. These indicators are related to figures

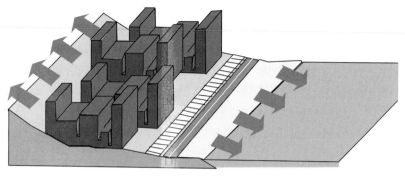

URBAN AREAS ALONG THE COAST

FIGURE 7.5.
Fragmentation effect caused by coastal linear urban areas.

of distribution, density, continuity and characters of urbanisation and have to consider the relationships among the different land uses in the environmental system, which can be outlined by research. The goal is to determine, via parameters, the presence of the real environmental continuity condition of single ambits through the monitoring of barriers and causes of environmental fragmentation due to human impacts:

The following indicators can be used:

- impact of urbanisation on the different land uses in relevant geographical ambits
- dispersion of the urbanised area (related to road network density)
- spatial character of settlement organisation
- infrastructural fragmentation (area and surface type of non-permeable infrastructures)
- agricultural fragmentation (forms of agricultural spaces, compactness and dispersion coefficients)

Big complex barriers can be linear or concentric. In Italy the latter are represented mainly by agricultural and productive settlements, which cause high disturbance to eco-connectivity and present a low level of biopermeability. Big complex barriers of a linear type are the multiple infrastructures, such as a combination of highway, railway and road on the same line (Spellerberg 1998). This linear type of barrier is very common in Italy and constitutes, with its physical boundaries, a total obstruction to environmental continuity (Figure 7.6). The ecological connectivity is here only possible when the infrastructures transit via a tunnel or viaduct (Figure 7.7). Because of their high impact on the territory, it is important to take into account these elements from the early stages of the research.

In the case of long tunnels the infrastructure does not produce disturbance. Viaducts and bridges, on the other hand, although allowing the transverse

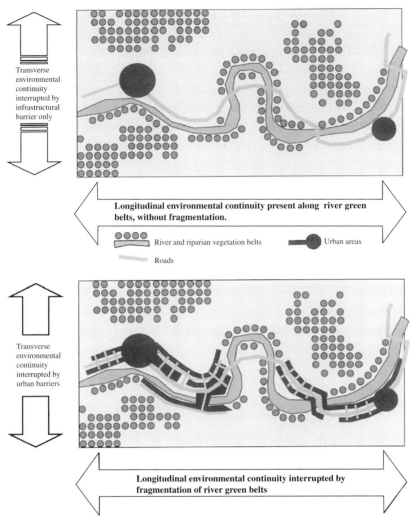

Transverse environmental continuity interrupted by infrastructural barrier only

Longitudinal environmental continuity present along river green belts, without fragmentation.

River and riparian vegetation belts

Roads

Urban areas

Transverse environmental continuity interrupted by urban barriers

Longitudinal environmental continuity interrupted by fragmentation of river green belts

FIGURE 7.6.
Effects of linear urbanisation on environmental continuity in a valley bottom.

FIGURE 7.7
Example of environmental
continuity in the
Central Apennines,
where road tunnels
allow ecological connections.

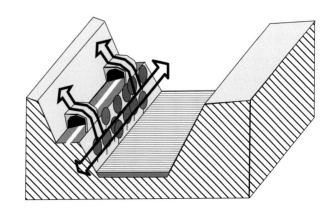

passage of species, cause relevant noise and vibrations. Past research has shown that ungulates generally remain 500 m away from these viaducts (Ministry of Transport 1996). The amount of disturbance has to be assessed for each single case, considering the length and the height of viaducts and bridges, and the general characteristics of the main infrastructure.

In conclusion, the evaluation of large complex barriers can reveal limited areas in Italy where environmental continuity can be developed or restored. More difficult is the opening of artificial ecological passages through the multiple infrastructures, owing to the complexity and high cost of the work, the low social awareness of its necessity, and consequently the political reluctance to approve it.

The barriers defined as 'simple' do not represent a rigid obstruction to environmental continuity. This is demonstrated by a large number of wildlife road casualties, especially small mammals, reptiles and amphibians. Traffic and noise are the main disturbance elements that condition the fragmentation degree related to the simple barrier. An example of a simple barrier is a road with only one carriageway.

In the case of viaducts and tunnels (Figures 7.7, 7.8) the considerations expressed before for big complex barriers are valid here too. For the analysis of single barriers, however, it is necessary to verify whether other elements of environmental fragmentation are present. An example in Italy are the little upland roads, where several conditions of physical occlusion to eco-connectivity can occur, e.g. supporting walls on road curves and slopes. In this type of barrier, monthly and hourly traffic fluxes can be recorded. This allows the temporal data on wildlife movements to be related to the disturbance levels, in order to evaluate the real needs of the defragmentation work.

Other types of simple barrier the high- and low-tension power lines and rural road systems. The latter are typical of intensive agricultural areas, where the disturbance factors are related to the agricultural activity.

FIGURE 7.8
Highway viaducts can often reduce fragmentation by infrastructure in the mountains.

7.7   Developing ecological networks

From a large-scale model it is possible to draw up the design of the Italian environmental continuity, characterised by the dominant role of the Alps and Apennines, by their significant buffer zones and other dispersed residual natural areas. This continuity is, however, only apparent because of the numerous already mentioned barriers to biopermeability, especially infrastructures and urban areas, which are extremely frequent, in Italy as in most European countries.

From the design of the national environmental continuity, it is then possible to move to the development of ecotones. A topic that requires further research, however, is the relationship between the environmental continuity structure and the ecological network that is single-species-oriented. A first type of complexity concerns the interaction between the environmental continuity system, with reference to the human components, and the design of ecological networks, with reference to the biological components. A second type of complexity concerns the problem of maintenance, and eventually restoration, of those conditions of environmental continuity that are fundamental for any policy on habitat defragmentation.

In order to overcome these complexities it is necessary to operate with the plan as an instrument at different levels. This brings about a new debate on the planning hierarchy. Interventions on the defragmentation of natural habitats can be seen for instance in alternative solutions to contiguity between green spaces, soil re-naturalisation, infrastructural by-passes and local agricultural policy objectives. These interventions can be successfully managed only at municipality and lower levels. But at these levels the strategic configurations of environmental continuity are neither evident nor assessable, while they function at national and regional levels. The latter, on the other hand, have great influence on the management of interventions at lower levels.

One of the answers that current research in Italy should provide is indeed to the problem of relations between different levels, scales, legislation, management and implementation in ecological networks. In Italy, accordingly, the bridge between the studies and the legislation becomes imperative. In this context, the system of environmental continuity could become the reference for any further planning and programming action of territorial transformations.

Studies conducted at national level (Romano 1999) show that the environmental fragmentation process is still active and the establishment of protected areas, though numerous, is not enough to stop such a trend. The answer instead can be found in large-scale and long-term policies, able to control interventions on infrastructure development, urban expansion and natural area consumption in order to conserve at least the present conditions for environmental continuity.

Research properly related to ecological networks according to different species (Battisti and Contoli 1997), with indication of home range and ecological corridors, is still limited to restricted areas and a long time has to pass before the knowledge will reach that geographical context able to support a proper design implementation. The maintenance of environmental continuity conditions, where possible, can represent clearly a key preamble with responsibility functions for a planning system aimed at setting up instruments able to retain ecologically and culturally sound programming inputs, inside and outside protected areas.

7.8   Discussion

At present in Italy there is a lack of ecological data at different scales useful for the design of ecological networks with different levels of detail. At the same time there is a gap among the scientific, social and political communities on the knowledge of environmental continuity and ecological needs of species and habitats (see Figure 7.9). Nevertheless, it is necessary to prevent urban processes that are increasing the fragmentation conditions present in the territory.

To add to this, the natural and cultural elements are here tightly interconnected and thus difficult to consider separately. Planning actions, having the capability to deal with both natural and cultural aspects, can directly contribute to the defragmentation process with the use of information available at the moment, even if ecological data are not extremely precise, as some authors have already suggested (Forman and Hersperger 1997). Moreover, while ecological research proceeds in studying the characteristics of the territory in terms of habitat distribution, core areas and ecological corridors, it is important that planning, in particular at the small scale, uses some precautions to avoid land fragmentation.

These precautions can be related to the following contents of the plan:

- spatial distribution of new urban areas, controlling large interruptions to the local environmental continuity
- control of building typologies, avoiding the use of low-density developments that have a large impact on the territory
- control of new road projects, inserting the environmental defragmentation elements of general type
- control of urban green areas, maintaining the physical connection among them and the other natural and semi-natural green areas surrounding the urban area

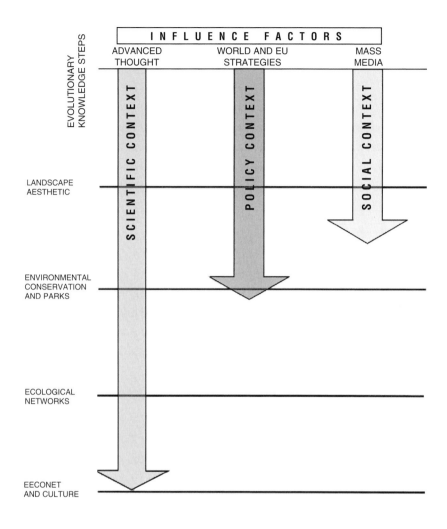

FIGURE 7.9.
Levels of knowledge on the relationship between ecology and culture in Italy.

In this way it is possible, at a general planning level, to support defragmentation conditions of areas with ancient human influence. When more ecological information on the geography of home ranges and spatial relations of the species is available, it will be possible to control the specific planning for the design of single elements and urban objects.

This is one of the main contributions that research on environmental continuity and biopermeability can offer to ecological connectivity in Italy. However, in areas of ancient tradition it is important that these concepts be theoretically applied considering the balance between nature and culture, and practically used according to the operative scale of the plans.

7.9   Conclusions

A methodology to assess the state of an ecological network has to take into account a certain number of criteria that should not be confined to species habitats, but should be applied to the larger context of landscape. The latter is becoming a fundamental topic in studies of nature conservation. In areas of ancient human usage it represents the idiosyncrasy of the cultural elements in an ecological network. The wider landscape, therefore, is the appropriate context in ecological network development. Knowledge of the cultural landscape, applied landscape ecology and environmental policy can assist in this process.

A common objective of European policies is to reinforce the ecological sustainability of the environment. This should be considered within the framework of sustainable development, which retains both natural and cultural aspects. Many studies, however, lack the latter. It is hence proposed to use the wider public policy context, including sustainable development, in ecological network development.

Another common objective of European policies is to conserve biological and landscape diversity. In order to achieve this it is necessary to move beyond ecology and consider the human influence on the ecological structure in terms of socio-economic needs and pressures on the territory. The European policies, accordingly, have to be supported by national, regional and local policies dealing with physical planning and having thus a strong cultural input.

In Italy, however, physical planning has met a series of limitations that have led us to advance the concept of the 'de-anthropisation of the plan'. The concept moves from the consideration that in the country any practical and implementation action on the defragmentation of natural habitats can be planned only at municipality level. But at this level the environmental continuity is invisible, whereas it is more evident at national and regional level. This is therefore the scale to use for theoretical inputs and actions.

Past research shows how ecological networks are the results of science based on nature conservation, landscape ecology, land use and environmental planning (Jongman 1995). This indicates how human activities, education and awareness play here a relevant role. Accordingly, the development of ecological networks in Europe and Italy should be carried out incorporating cultural aspects into the features of an ecological network and, in turn, considering the interaction between its natural and cultural components. Socio-economic trends and functions of the site should be taken into account, linking ecology to environmental economics. Then, the integration of ecological networks into the planning system becomes imperative.

In conclusion, the wider landscape context should be used in ecological network development, including the human processes that shaped the landscape and its natural and cultural components. The wider public policy context can be addressed in parallel, with the goal of balancing nature conservation with sustainable development. This assumption emphasises how ecological networks, in areas of ancient occupation like Italy, are located between nature and culture. For this reason the planning of future landscapes in these areas should include not only ecological and natural aspects, but also cultural, social, economic and political considerations.

MARIUS BOLCK, GIUSEPPE DE TOGNI, THEO VAN DER SLUIS
AND ROB H. G. JONGMAN

8

From models to reality: design and implementation process

8.1   Introduction

The concepts and models that have been developed and that have been
explained in the preceding chapters should find their way into practice. The
importance of models, adequate data and the translation into practice, in both
design and implementation by organisations responsible for it, is pivotal for
a new and improved landscape planning and nature conservation strategy. In
Europe many authorities at local, regional and national level have worked out
plans for ecological networks (Jongman and Kristiansen 2001). In different settings and with different budgets and constraints all governments work in the
same direction of a more sustainable landscape and maintenance of natural
ecosystems.

In this chapter we will show and compare the approaches in two different
regions of Europe, Bologna and Modena (Northern Italy) and Gelderland (the
Netherlands) using the same models and having comparable approaches in different political settings. Both regions develop ecological networks but are in a
different phase and have different possibilities (Boxes 8.1 and 8.2).

The region and provinces involved have a problem in maintaining and improving biological diversity that they want to solve through the design of an
ecological network. Owing to the different situations different plans are made,
but with the same methodology. The general plan is made at the provincial
level and plans are specified at the local level. With a practical plan that gives
broad views and is worked out at the local level, it is expected that concrete steps
can be set in contact with municipalities and district water boards for realising
local plans and actions to improve or make new corridors.

In the Netherlands the ministry of Agriculture, Nature Management and
Fisheries already published a National Ecological Network in 1990, whereas
in Italy an exploration process for new nature conservation strategies started in

BOX 8.1. The provinces of Bologna and Modena

The provinces of Bologna and Modena stretch from the southern part of the Pianura Padana to the northern part of the Apennines. The plain forms the northern part while the southern part is occupied by the mountain range of the Apennines. The Via Emilia – a road of Roman origin – runs along the foothill. The major cities are developed along it.

The mountains are characterised by the dominance of natural elements, whereas the plain is dominated by intensive agriculture and urban areas. The mountain ecosystems reflect an appreciable nature and landscape. The landscape in the plain is poor in biodiversity. The main rivers still act as a natural refuge – thanks to grass dykes, and some small wet woodlands within the floodplains – and as the basic landscape structure. Moreover, the plain territory hosts few cultural and partly natural elements: old renaturalised hemp maceration pools, relicts of hedges and tree lines, isolated big trees and historical parks of villas and farm buildings. All in all natural habitat forms less than 5% of the total surface area of the plains of the provinces of Modena and Bologna.

During the last few years the areas in the past dedicated to industrial agriculture have been increasingly transformed to nature through rehabilitation financed by the European Commission Agricultural and Environmental Policy funds. Wetlands, grasslands, small woods, hedges and tree lines have been created.

The foothill area connecting the hills with the plains is most congested. It is dominated by the Via Emilia and the motorway, and the high-speed railway (under construction) will run through it. Besides, it is the zone with the greatest urban and industrial settlement concentration. The rivers and canals form the major linkage between the hills and the plains area and they cross many barriers, including the motorway and the high-speed railway.

The distribution of the fauna in the territory depends on the ecological features, either natural or defined by humans. The zones important for the wild fauna are in particular in the hilly–mountainous part of the territory (outside the study area), since the plain offers only few relicts and a patchwork of new rehabilitation areas. In spite of that, the plain is the very place where it has been possible to record both the increase and diversification of the fauna, thanks to the habitats restored. For example, during the past 10 years, colonisation of several new species was recorded: either spontaneous (cattle egret, red-footed falcon) or due to people (greylag goose, Egyptian goose) and recolonisation of species (grey heron, squacco heron, little egret, great white egret, bittern, spoonbill, mute swan, pochard, shoveler, cormorant, black-tailed godwit).

BOX 8.2. The Province of Gelderland, the Netherlands

The province of Gelderland is situated in the centre of the Netherlands and is characterised by three major landscapes. In the south, the river Rhine dominates the landscape. In the northwest the morainic hills of the Veluwe, remnants of the Saalian ice age, are situated and in the eastern part Pleistocene sands intermingle with lowland brook systems.

As elsewhere in the Netherlands land use is intensive and population is dense (382 per km^2). Areas of unspoilt countryside are becoming increasingly fragmented and isolated as a result of building roads and houses, and because of intensive farming. The province is an important traffic corridor over road, rail and water between the western part of the Netherlands (Amsterdam, Rotterdam) and Germany.

Gelderland is relatively rich in species, having large more or less natural areas in its territory. It has a rich flora, partly characteristic of the poor sandy soils of the north and east part of the province, partly riverine flora and fauna transported by the Rhine from the German and Swiss hinterland.

The Dutch government is devoting considerable effort to creating a network of protected areas throughout the Netherlands: the National Ecological Network. Even after addition of new conservation areas, however, the habitat is still often not large enough for flora and fauna to survive on a sustainable basis. The biological diversity in the rural environment of Gelderland can be increased again if there are connections between the various nature conservation areas. These ecological corridors are the backbone of the wildlife of the Gelderland countryside and thus form a green network for plants and animals. For many animals it is essential to create crossings (ecoducts or tunnels) to cross obstacles such as motorways.

Since 1996, the Green Connections project has been developing thirty-five ecological corridors that have to be created in the province by 2020. The aim is an integrated approach, linking the project with rural renewal. While the province has overall responsibility for creating ecological corridors, the municipalities and the water boards carry out project co-ordination. The development of nature conservation areas and corridors focuses primarily on the needs of target species. The province has developed models for the conditions that an ecological corridor has to satisfy for these species. Many species of flora and fauna that live in the same habitat will also benefit.

An ecological corridor runs past villages, across fields and meadows and requires the consent and goodwill of the landowners and the people who use the land. Political support at all levels of government is also indispensable, as is the co-operation of local authorities who have been asked to include the proposed ecological corridors in their zoning plans. Two years of work on three pilot projects in Gelderland resulted in implementation plans supported by all involved parties.

1995 on the initiative of the national environmental planning agency (ANPA). In the Netherlands, decentralisation of nature conservation from national to provincial level took place in the period 1993–6, whereas in Italy all four government levels (national, regional, provincial and local) have their own responsibilities in this field. In the Netherlands the provinces elaborate the National Ecological Network (NEN). They designate nature reserves and develop nature restoration projects and ecological corridors based on this indicative national plan. In Italy a number of regions and provinces together are trying to develop new ways for conserving nature through ecological networks. The Province of Gelderland started in 1996 with the project 'Green Connections', with the aim of realising ecological corridors. With the same aim in 1998 the Province of Bologna and in 1999 the Province of Modena started a similar project.

In the human-dominated landscape, many barriers and gaps exist that have a negative effect on species movements. Roads, highways and recreational trails cause gaps in the natural habitats and have a negative impact on wildlife by traffic noise and lights. Reducing the effects of barriers and reducing traffic impact on species movements is one of the aspects of corridor development in land use planning (see also chapter 5). In developing mitigation and compensation measures (i.e. tunnels and ecoducts), species-specific requirements are an important consideration in the design.

Impact reduction of traffic and recreational or other users of the landscape is a second design principle in the multi-functional landscape. It is important to reduce the influence (negative impact) that other users of the landscape can have on animals using ecological corridors. Another aspect to consider in the design and management of ecological corridors is time. Ecological corridors, once they are established, do not remain constant in time. Succession and changing environmental circumstances result in changing vegetation. Some vegetation types are characteristic of early stages, whereas others, such as large (tall) trees, develop only in later stages, sometimes only after long periods of time. The different characteristics of an ecological corridor in time may affect their function for specific animal species. That means that ecological corridors require management.

This chapter first discusses the way in which available models have been applied, and then shows in detail how the design and implementation process develops at regional and local levels in two European regions.

8.2   Modelling: the LARCH analysis

An ecological network should be geared towards an ecosystem (forest, marshland, moors) or species. A strategic choice of target species benefits many more species than an arbitrary sole species in the network design (van der Sluis and Chardon 2001). There are focal species that have broad-scale effects at the ecosystem level (Simberloff 1998; Dale *et al.* 2000): keystone species (top predators, e.g. wolf, brown bear, otter), ecological engineers (beaver, red deer) and umbrella species (red deer). Umbrella species either have large habitat requirements or use multiple habitats and thus overlap habitat of many other species (see also Groot Bruinderink *et al.* 2001, 2003).

The landscape-ecological model LARCH (see chapter 4) was used to assess the ecological networks under the present situation and a development scenario in the provinces of Bologna, Modena and Gelderland. With LARCH one can (1) identify the functional ecological network at present, (2) assess whether the designed ecological network will result in an improvement of the present situation, and (3) identify opportunities to optimise the ecological network.

The area or ecosystem is assessed on presence of habitat for the selected target species. Based on the quality and quantity of habitat it is defined what potential populations are, and whether these populations can be considered viable. A population is considered sustainable or persistent if the chances of extinction are less than 5% in 100 years (Shaffer 1981).

For the provinces of Bologna and Modena three ecosystem types were selected, which cover most important natural habitat types in the study area: woodland, wetland, and grassland. To assess whether these ecosystem types might function for specific wildlife species, species were selected which can be considered representative for these ecosystems (Table 8.1). The selected species operate on a scale that is appropriate for this landscape, with different dispersal ranges and some also sensitive to barriers. For these species, it was assessed whether the ecosystem still functions as an ecological network.

It has been verified, with the help of LARCH and LARCH-scan, what the present connectivity and fragmentation is in the provinces of Bologna and Modena. An overview has been given of the present situation for species and populations involved, and new scenarios have been developed. The result of the analysis shows that the two provinces have a serious fragmentation problem. Obviously, the remaining area with natural habitat (only 5%) (van der Sluis *et al.* 2001) is too small for many species present. The natural areas can only partly function as a network: many species suffer from fragmentation. In the scenarios for future development, shapes of the new corridors have been included (strips of land with edges, trees, grass and ponds) as a 'generic' project in order to test the possible connectivity of the scenario.

Table 8.1. *Selected species for analysis with LARCH for the provinces of Bologna and Modena*

Species sensitive to barriers are shown in bold type.

Habitat type	Barrier sensitivity	Small range (0–10 km)	Large range (10–50 km)
Woodland	**sensitive**	–	**European polecat (Putorius putorius)**
	not sensitive	red-backed shrike (*Lanius collurio*)	turtle dove (*Streptopelia turtur*)
Wetlands/ marshland	**sensitive**	**Italian crested newt (Triturus carnifex)**	–
	not sensitive	banded demoiselle (*Calopteryx splendens*)	bittern (*Botaurus stellaris*)
Grassland	**sensitive**	–	–
	not sensitive	stonechat (*Saxicola torquata*)	yellow wagtail (*Motacilla flava*) quail (*Coturnix coturnix*)

Source: van der Sluis *et al.* (2001).

One of the main conclusions is that habitat requirements for most selected species are high. With realisation of the scenario, some species (bittern, stonechat, red-backed shrike) will still be under threat, despite the ambitious scenario. For most species the spatial cohesion has improved greatly,
resulting in formerly isolated areas being linked into the larger ecological network (Figure 8.1).

Fragmentation also appeared to be an important factor causing the decline
of species in Gelderland. Here also the question was which species and which
habitats should be linked and why. In Gelderland 30 focal species have been
selected as being representative of different ecosystems (forest, heathland,
grassland and water/marsh). Species differ in dispersal capacity and dispersal
distance. Among the selected species are pine marten (*Martes martes*), tawny
pipit (*Anthus campestris*), bittern (*Botaurus stellaris*), great crested newt (*Triturus
cristatus*), large copper butterfly (*Lycaena dispar*) and brook lamprey (*Lampetra
planeri*).

Based on presence of suitable areas and barriers (roads) the persistence and
the isolation of populations has been calculated and spatial cohesion of the network has been defined (Figure 8.2). In this way, the functioning of designed

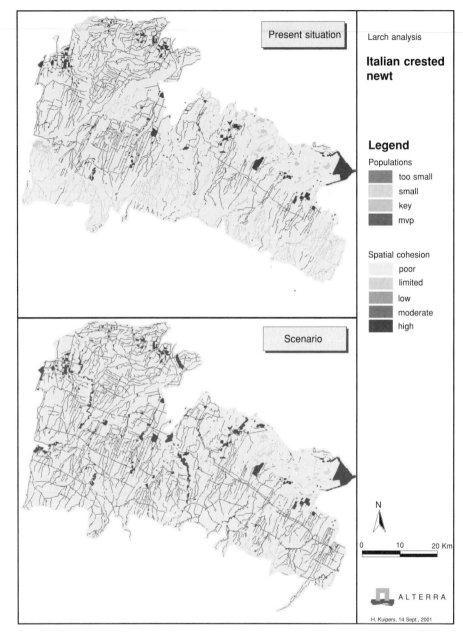

FIGURE 8.1.
Existing networks of the Italian crested newt, showing minimum viable populations (mvp) in the present situation and in a possible scenario for the provinces of Bologna and Modena (van der Sluis *et al.* 2001).

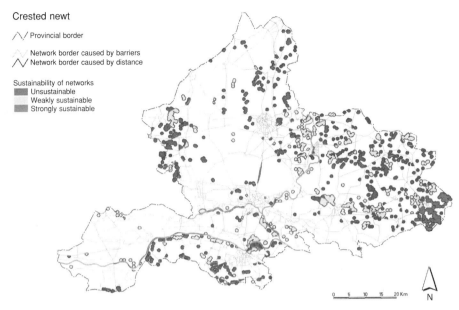

Crested newt

⁄\⁄ Provincial border

Network border caused by barriers
⁄\⁄ Network border caused by distance

Sustainability of networks
▓ Unsustainable
▒ Weakly sustainable
▓ Strongly sustainable

FIGURE 8.2.
Networks of the great crested newt (*Triturus cristatus*) are very fragmented because
of its short dispersal distances and the density of roads in Gelderland. Many
populations are not sustainable in the long term. Linkage can improve the
sustainability of the populations (Koolstra and Reijnen 1998).

networks has been assessed. In addition, bottlenecks for sustainability can be
calculated and solved (Figure 8.3).

The scenarios developed in Gelderland, Bologna, and Modena have a relatively large impact on the woodland ecosystems. Under the proposed scenario for the provinces of Bologna and Modena the total woodland habitat
increases by 40% up to 8180 ha. In the new situation species dependent on
woodland habitat are still limited owing to lack of habitat. For marshland the
development scenario results in better spatial cohesion. Only a few smaller
areas do not form part of the ecological network (Figure 8.4). Large differences occur for the species analysed. For the Italian crested newt (Figure 8.1)
the number of local populations has decreased by 50% but they are merged
into a more stable Minimum Viable Population (MVP). The population is
sustainable both at present and after implementation of the development
scenario.

For the bittern, the lack of extensive marshland habitat is currently the main
bottleneck. Fragmentation is less of a problem for this species, owing to its
large dispersal capacity. In this scenario persistence improves greatly, but the
population is still dependent on immigration from other areas. The results

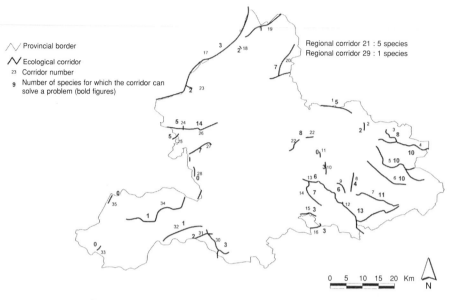

FIGURE 8.3.
Number of species for which an ecological corridor can solve a problem of
sustainability according to the LARCH model applied to the province of Gelderland.
The lines with the small numbers indicate potential corridors. The bold numbers
indicate the number of species (Provincie Gelderland 1999).

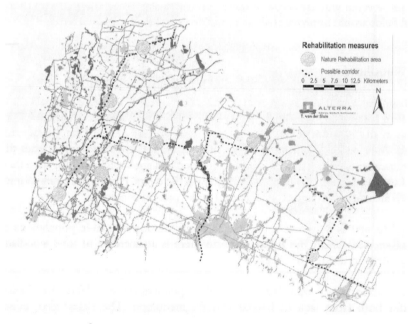

FIGURE 8.4.
Possible interventions to improve the ecological network of Bologna and Modena
(van der Sluis *et al.* 2001).

emphasise that considerable efforts have to be made to improve the situation substantially for the large herons in this area. The stonechat in the area of Bologna and Modena shows a slight improvement under the development scenario, owing to increase of available habitat and improved spatial cohesion of the ecological network.

8.3   Towards sustainable networks

8.3.1 The function of the ecological network

Biological knowledge and modelling results must be translated into technical solutions and policy. Design and management of linkages for conservation can be viewed in a biological way, in a socio-political way, and as a design problem (Bennett 1999).

The biological purpose of an ecological network and how it is intended to provide benefits to flora and fauna is an important first step and an essential basis for evaluating design and management of the landscape. Within an ecological network, ecological corridors are species-specific and they can have a variety of functions. Knowledge of the ecological structure and processes in the landscape, combined with the behaviour and ecology of species, is of utmost importance in the design of ecological networks and corridors. In all cases the landscape has to become accessible for species using hedgerows, streams and small forests for guidance and shelter.

In Modena and Bologna the general approach has been to enforce the existing natural structures. Despite the scarcity of natural features the plain still presents linear and larger-sized elements. The larger elements are concentrated in the northern part of the territory. The linear elements correspond to main rivers and canals crossing the plain from south to north towards either the Po River or the sea. They connect landscape elements and in this way they form the backbone of the ecological network. Based on this, the provincial ecological network layout has been designed based on two main principles.

1. To strengthen and improve the north–south corridors, with a particular interest in the foothill zone, which acts as an important ecotone. 'To strengthen' means identifying areas suitable to be renaturalised (creation of entire new ecosystems); recognising strategic elements bridging existing natural and semi-natural areas; widening those corridors with natural vegetation (e.g. wetland forest, grass strips); and increasing the number of natural habitats important for the ecological network structure (core areas

and corridors). 'To improve' means to preserve and manage with nature restoration techniques the existing landscape and natural features regarded as biodiversity sources and ecological corridors (adding natural features to an existing ecosystem).

2. To strengthen the east–west line. This aims at adding new transverse corridors between the rivers and canals, i.e. the main natural structures in the plain area.

The existing features, such as rivers, canals, wetlands, hedges, etc. are regarded as starting points for the plan. Other – partly new – elements have been included as core areas or corridors. These are at times multi-functional areas, like floodplains, which are excavated to allow creation of retention basins for the purpose of flood protection, resulting in marshland areas. In the resulting ecological network (Figure 8.5) the position of corridors (dotted lines ending with arrows) depends on a specific local analysis, directly involving municipalities and other stakeholders interested in the future realisation. The design of a network consisting of linear elements and natural areas, connected as much as possible, is based on different needs: protection, preservation and creation of natural habitats; landscape improvement and expression; and defining greenways for recreation. There is a strong consciousness of the necessity of integrating the ecological network with other social functions in order to involve various interests, build public support and to acquire funding.

8.3.2 Species-related landscape models

The conclusion from the modelling for Gelderland was that ecological corridors are mainly of importance for medium-sized species with good dispersal capacity. Many smaller organisms have restricted dispersal capacity and can only maintain small populations. The main conservation strategy for these species is enlarging and improving their habitat. Larger mammals and birds have good dispersal capacity but also need larger habitat areas.

Having stated how much and what type of habitat should be available, it is still not clear to most of us what the landscape will be. To understand and communicate these ideas, one needs landscape models. This is a way to summarise and make visible what focal species and habitat requirements mean. The province of Gelderland did summarise the multitude of possible focal species in seven landscape models named after focal species and focusing on linkage of specific habitats (Table 8.2, Box 8.3). Many more animal and plant species are expected to make use of this kind of landscape structure. The models consist of a zone 250–500 m wide with a continuous corridor and/or stepping stones.

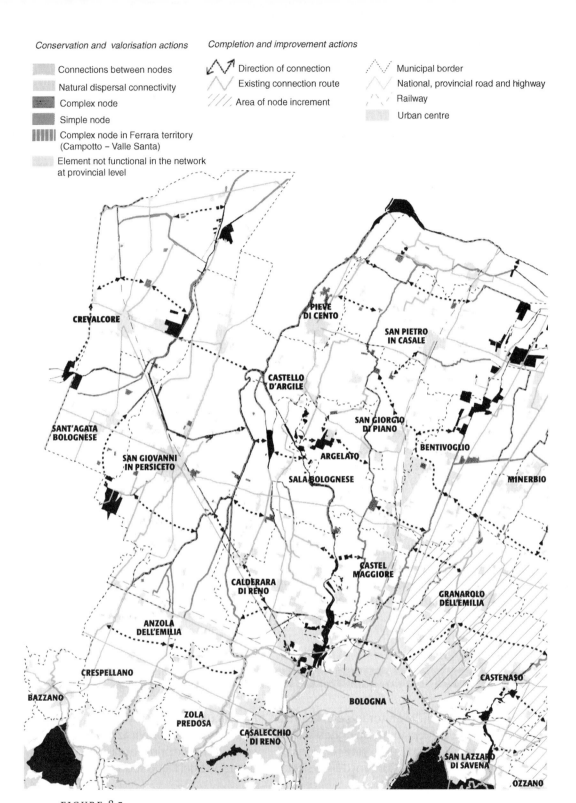

CREVALCORE

PIEVE
DI CENTO

SAN PIETRO
IN CASALE

CASTELLO
D'ARGILE

SANT'AGATA
BOLOGNESE

SAN GIOVANNI
IN PERSICETO

SAN GIORGIO
DI PIANO

ARGELATO

BENTIVOGLIO

SALA BOLOGNESE

MINERBIO

CASTEL
MAGGIORE

CALDERARA
DI RENO

GRANAROLO
DELL'EMILIA

ANZOLA
DELL'EMILIA

CRESPELLANO

CASTENASO

BAZZANO

BOLOGNA

ZOLA
PREDOSA

CASALECCHIO
DI RENO

SAN LAZZARO
DI SAVENA

OZZANO

FIGURE 8.5.
The ecological network for the province of Bologna (Emilia-Romagna, Italy). The
larger shaded areas in the south and north represent the core areas. The streams
between them are the main ecological corridors. Smaller corridors (arrows) connect
the streams. Courtesy Centro Agricoltura Ambiente S.r.l.

Table 8.2. *Overview of the landscape models for ecological corridors developed by the province of Gelderland and based on habitat requirements and landscape structure*

Landscape model	Focal species/habitat	Characteristics of the landscape zone
badger (*Meles meles*)	intermediate and larger mammals	wooded banks, small forests (8%), 500 m wide
great crested newt (*Triturus cristatus*)	amphibians	corridor and stepping stones, 250 m wide
lizard	reptiles, butterflies	corridor and stepping stones (1 and 10 ha) with oligotrophic grassland or heathland, 250 m wide
copper (*Lycaena phleas*)	butterflies	stepping stones (0.5 and 4 ha), oligotrophic grassland or heathland, 250 m wide
white admiral (*Limenites camilla*)	butterflies	stepping stones, well structured landscape, humid forest, 250 m wide
sedge warbler (*Acrocephalus schoenobaenus*)	reed birds	stepping stones (2.5 and 25 ha), reed marsh
ide (*Leuciscus idus*)	brooks, streams	natural banks, spawning places

Source: Koolstra and Reijnen (1998).

For walking species with a high dispersal capacity, a corridor for movement and foraging might be sufficient in most cases on the scale of regional planning. This is not necessarily so for plants and smaller walking and creeping species. Flying species can make use of high-quality steppingstones for foraging and reproduction. Some intermediate species need a kind of string of beads, a combination of connectivity through physical ecological corridors and larger steppingstones as temporary habitat.

For each ecological corridor several landscape models can be applied. Minimum habitat requirements have been described and a design in a bird's-eye view illustrates how this can appear in reality (Figures 8.6, 8.7 and 8.8).

BOX 8.3. The 'Crested newt' Landscape Model

The model 'Crested newt' consists of an ecological corridor with stepping stones embedded in the landscape. Ponds and other wet elements are essential. Other species are also supposed to profit from this corridor: the moor frog (*Rana arvalis*), the tree frog (*Hyla arborea*), the common spadefoot (*Pelobates fuscus*) and the grass snake (*Natrix natrix*).

The model aims for the restoration of a small-scale landscape including wet biotopes. Depending on the abiotic basis (river clay, sand) the practical outcome of this model will differ considerably. It offers habitat for a great diversity of species from shrew to little owl and dragonflies to bats. A conflict can occur if fish is put out in the ponds because that diminishes the survival chances for young frogs and toads.

Landscape zone (1.2 ha km^{-1})
Small-scale landscape with a coherent network of linear elements in the cultural landscape. Additionally at least five small ponds of about 500 m^2 are found in every kilometre length of the network. The width of the landscape is 250 m.

Corridor (1 ha km^{-1})

Vegetation:	Shrub, humid oligotrophic grassland, woodlands (deciduous), wooded banks, drains, ditches, brooks and their banks.
Minimum width:	10–15 m
Maximum length:	500 m
Maximum interruption:	50–100 m
Barriers:	Roads, railroads and urbanisation. Fields have a smaller barrier effect. Mitigating measures are needed such as the use of tunnels. Bigger tunnels (diameter > 1 m) are more effective.

Stepping stone (3 ha km^{-1})

Vegetation:	Pond with well-developed water and bank vegetation and open spaces; terrestrial habitat consists of shrubs, hedges or wooded banks with sufficient dead wood and holes as hiding places.
Minimum surface area:	1–3 ha terrestrial habitat sites with some larger ponds situated in it (about 2000 m^2).
In-between distance:	1 km

FIGURE 8.6.
Bird's-eye view of the model 'Crested newt', characterised by hedgerows, small wetlands and ponds.

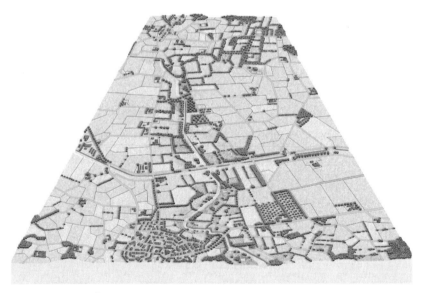

FIGURE 8.7.
Bird's-eye view of the 'Badger' model, characterised by small forests, hedgerows and grassland.

8.3.3 Landscape design

For water-related species, natural stream banks must be maintained, and where roads cross streams and canals, tunnels have to consist of both a dry and a wet passage for fauna. The design of an overpass over a motorway for

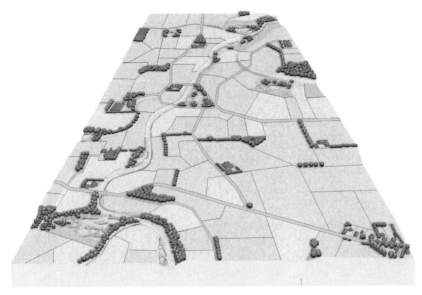

FIGURE 8.8.
Bird's-eye view of the 'Ide' model, dominated by a stream and stream-related wetlands.

red deer requires a mosaic-like landscape, whereas badgers prefer hedgerows. Within the landscape 'Badger' model details of road crossings can be worked out in design plans (Figure 8.9) as well as in practice (Figure 8.10).

For some wetland species and woodland species in Bologna and Modena there is definitely a need for large core areas to be developed to improve the size of the population. These areas should be located near other larger areas. Strategically planned forest and wetlands could improve sustainability and corridors, and some woodland could be strategically located as stepping stones between existing reserves. Ecological corridors are built upon the existing landscape pattern. This means that landscape elements present are strengthened, and landscape elements might be developed in line with existing structures, to improve the network and spatial cohesion.

In the ecological network as designed by the provinces of Bologna and Modena no clear assumptions have yet been made on the design of the corridors. Design criteria and landscape dimensions are needed. Owing to the character of the landscape, the species that occur in it and meetings with stakeholders, the following design principles have been developed.

1. Main existing corridors are formed by the watercourses, rivers and streams, and should be strengthened. Ideally the corridor is formed by the river and is lined with wooded banks, with trees up to 10–15 m high and some 6 m wide, with undergrowth or shrubs on one side (Figure 8.11). Within the river

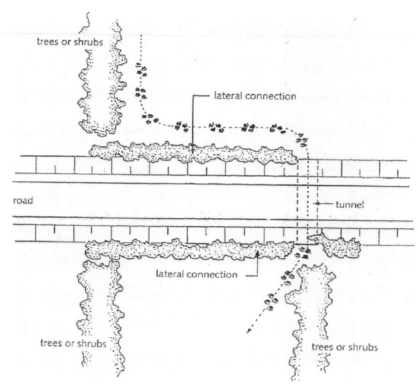

FIGURE 8.9.
Details of a landscape design adapted to badger requirements. The road is a barrier and is crossed by a tunnel. Wooded banks lead the badger to and from the tunnel. The road is fenced to prevent traffic collisions.

embankment areas wooded banks can be located along the water. Where this is not possible, owing to other forms of land use, the width of the floodplain or management requirements from the water boards and these wooded banks should be realised outside the embankment areas.

2. Transverse corridors might be a strip 10 m wide consisting of shrubs and grassland. The vegetation should for all corridors consist of natural vegetation, indigenous species, preferably following spontaneous germination and settlement of species.

The land required to maintain or establish linkages may be in private ownership or public (government) ownership, or it may comprise multiple parcels with a diverse range of owners including private individuals, companies, government agencies or authorities and community or conservation groups. Ideally, a long-term arrangement with the responsible land managers is required to ensure that there is an ongoing commitment to the objectives of the linkage (Bennett 1999).

FIGURE 8.10.
A newly constructed badger tunnel near under the A73 motorway near Nijmegen, the Netherlands.

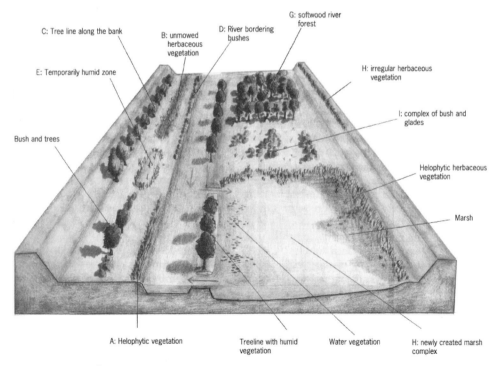

FIGURE 8.11.
Nature restoration along the river corridors as proposed in Emilia-Romagna (Morisi 2001, Courtesy Centro Agricoltura Ambiente S.r.l).

In summary, it can be stated that design principles have to be given to assure successful realisation and functioning, as follows:

- minimal width: the wider the corridor the better it will serve multi-functionality; sizes have been given for different situations from 15 m to 200 and even 600 or 1000 m wide, varying for urban and rural situations and terrestrial and riverine corridors
- well established connectivity for species and humans, depending on the longitudinal design of the corridor and the barriers in it
- differences in use should be taken into account: people are using trails during the day, badgers move during dusk and dawn, amphibians migrate in the early spring and linkages should be adapted to that
- habitat diversity within the ecological corridor: a greater variety makes it more attractive for different species as well as for people
- accessibility from the surrounding land makes the ecological corridor multi-functional

8.4   The implementation process

Ecological networks are mostly designed at the regional level, but for implementation an in-depth analysis is required at the local level. The ecological network is important for the dispersal of plants and animals within the urban and agricultural fabric. It is necessary to approach the study from a local point of view in order to relate the interventions to a particular species. Besides, it is necessary to involve the municipalities because each new natural element will occupy part of their territory and the interventions will require their contribution for management and maintenance. That means that a bottom-up process must be followed, based on voluntary agreement and an active participation of the interested municipalities.

The experiences in the design of the ecological network by the provinces of Bologna, Modena and Gelderland show that work has to be done at different reference scales. The provincial scale is needed for a coherent design – especially for some target species – and the local scale allows the translation of the master plan into concrete projects. For Bologna and Modena the scenario foresees an increase of natural habitat by 2917 ha, mainly based on floodplains that are converted into wetlands and woodland (Table 8.3).

In the provinces of Bologna and Modena the ecological network has been implemented at a local scale until now in two areas: the Persicetano (Morisi 2001) and Bentivoglio. The ecological network functions merely as a vehicle to achieve restoration of ecological conditions and improvement of the landscape, and to give biodiversity some chances to survive. It is clear that planning

Table 8.3. *New habitat under the proposed development scenario for Bologna and Modena*

Type	Total area (ha)	Main land use (current)	Development: new vegetation type
East–west corridors	385.3	80% fields	50% meadow, 50% wet shrub
North–south corridors	650.8	84% water courses	50% wet shrub, 50% broad-leaved forest
Floodplains	1029.9	100% fields	50% wetland, 25% wet shrub, 25% broad-leaved forest
Other nature rehabilitation	851.6	89% fields	75% broad-leaved forest, 25% wet forest

an ecological network is just the first step in restoring the ecological conditions. The process starts with planning and includes maintenance over time, monitoring results, and dissemination of results. In Persicetano, during the planning phase, each core area and corridor has been verified, pointing out whether they were a new or an existing area. Moreover, within the new areas, 50 new landscape elements have been studied in detail in order to assess their realisation (Figure 8.12).

The Province of Bologna is preparing agreements with the municipalities involved to co-finance part of the projects. At the moment stakeholders and landowners are contacted for their co-operation in landscape restoration measurements (mainly planting of hedges or trees). In the process of implementation, political choices for a short or medium-term implementation programme will have to be made that can be discussed with municipality administrations and all other stakeholders. This is the crucial phase for the ecological network, because it should result in a co-ordinated plan for realisation and implementation.

To boost the implementation of ecological corridors the province of Gelderland started the Green Connections project. Realisation of ecological corridors requires an integrated approach because of the interaction with other land use claims such as agriculture, water management, infrastructure, housing and recreation. The province has asked water boards – for the water-based corridors – and municipalities – for the terrestrial corridors – to play a co-ordinating role in the implementation process.

The majority of the municipalities reacted positively and started the implementation process. Water boards have accepted responsibility for the aquatic

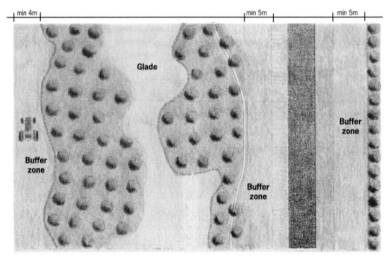

FIGURE 8.12.
Design examples of ecological corridors along a stream in the Po plain of the
community of Persicetano (courtesy Centro Agricoltura Ambiente S.r.l).

corridors. For every ecological corridor an implementation plan will be made
in which ecological and spatial aspects are combined, and in which a combination of functions (water management, ecological connectivity), responsibilities
and finances are worked out. Nature conservation organisations – responsible
for management of nature reserves – are supportive of this way of planning; agricultural organisations were more reluctant because of potential economic and spatial consequences. In discussion with them the landscape models
(Figures 8.6, 8.7, 8.8) appeared very instructive and gave a better insight into
the possibilities and problems. Farmers started to construct ponds and wooded
banks, and local agricultural environmental associations are active in realisation and management of landscape elements in the ecological corridor. In
The Netherlands, restricted possibilities do exist for financing the realisation
(acquisition and management). This regulation can be applied in the 35 designated landscape zones in Gelderland. The existing 161 500 ha of nature will
be expanded by 34 000 ha, of which 3 600 ha will be in corridors. Landowners
can apply for a subsidy for realisation of elements in the corridor. The results

Table 8.4. *Planned and realised new habitat in the province of Gelderland (status 2001)*

Planned new habitat in Gelderland	Planned	Realised
Proposed new nature development	20 000 ha	25%
Agricultural nature management	14 000 ha	50%
Total area of new nature	34 000 ha	35%
Ecological corridors in Gelderland		
Number of corridors	36	
Total length	935 km	
Total area of new habitat	3600 ha	
Stepping stones	47%	
Landscape elements	53%	

are good and various projects have been realised or are under construction (Table 8.4).

In the Netherlands it is striking that public support is not the greatest bottleneck in the realisation of ecological corridors. More problems have to be solved to make subsidy regulations applicable and to ensure the availability of land that has been allowed for at national level. Not all costs can be paid through subsidies, and the budget for road passages is insufficient. Ecological aspects are now usually included in road construction works, but not all existing problems can be solved in that way.

8.5   Conclusions

The two cases in Italy and the Netherlands show that it is possible to implement ecological networks. However, the process of implementation is complex. Medium-sized species appear to benefit most in these landscapes that are dominated by agricultural and urban land use. Communication with stakeholders is important and ecological considerations should be translated clearly to make them understandable to the wider public, particularly in their social and economic impacts.

It appears that in both regions the public support is bigger than was expected beforehand. Municipalities, water boards, public organisations, grassroots groups and agricultural organisations are willing to consider participation if the consequences are clear. Still, actual realisation in the field is not an easy task. The biggest problem is the finances for the projects.

The implementation process also requires enthusiasm, endurance, feeling for sensitivities, and skilled process management.

This process also taught us that much could be learned from European exchanges. It is important to exchange and share experiences and disseminate results. In addition, transdisciplinary research programmes are important to understand and realise ecological networks in practice. Ecologists, social scientists and land use planners, in close co-operation with politicians and land managers, can bring new perspectives to life.

9

Design of the Pan-European Ecological Network: a national level attempt

9.1   Introduction

The Pan-European Ecological Network (PEEN) is intended as a coherent assemblage of areas representing the natural and semi-natural landscape elements that need to be conserved, managed or, where appropriate, enriched or restored in order to ensure the favourable conservation status of the ecosystems, habitats, species and landscapes of European importance across their traditional range (Bennett 1998).

The ecological network model has been under development in Europe as a practical conservation tool for more than a decade. The concept of ecological networks is gradually becoming more and more important both in policies and practices of nature conservation throughout Europe. The proposal to establish a Pan-European Ecological Network – PEEN – by 2005 has been endorsed by ministers from 54 countries of the United Nations Economic Commission for Europe (UNECE). The network will be one of the principal means through which the Pan-European Biological and Landscape Diversity Strategy (PEBLDS) is to maintain and enhance the natural diversity of the continent (The Pan-European Biological and Landscape Diversity Strategy 1996; Rientjes and Drucker 1996; Bennett 1998; Külvik 1998; Work Programme... 2001).

The work programme for developing the PEEN is the first attempt at applying the network model on the continental scale. It provides the vehicle for pooling experience from existing networks at international, regional and national levels. The novel aspects of the approach are:

- geographical scope: embraces the whole UNECE region
- landscape dimension included through the Action Theme (AT) 4 of the PEBLDS

The principles on which the PEEN is based are the same as for the whole PEBLDS:

Building on existing agreements and initiatives
This aims at using all other existing international and national legislation and policies as well as facilitating, supporting and promoting the implementation of existing international agreements and treaties. This includes:

- the Bern Convention with the Emerald Network
- the EU Habitat and Species Directive forming the NATURA 2000 network
- Biosphere Reserves
- Important Bird Areas (IBA)
- Important Plant Areas (IPA)
- the Large Carnivore Initiative

Building on existing experience means *inter alia* that criteria and methods used by other agreements and initiatives will be utilised as far as possible for approaching the goals of the PEEN. The relevant project on formulating the criteria and methods for developing the PEEN is currently under way. The project on guidelines for application of existing agreements and initiatives is another high-priority project being implemented.

Integrating ecological considerations into relevant socio-economic sectors.
As ecological networks do not aim to provide total conservation, the conservation measures can often meet, at times unexpected, synergistic responses in sectors such as land-use planning, regional development, agriculture, forestry, fisheries, transport and tourism. The development of an integrated policy for establishing the PEEN is the subject of a separate project foreseen in the work programme.

One of the main objectives of the development programme for PEEN is to ensure that, as far as possible and appropriate, data, methodologies, evaluation techniques and other relevant information are available for the actors involved in developing and implementing the network. This information can be divided into two broad categories (Bennett 1998):

- information on the distribution and status of biodiversity of European importance
- information necessary to design and implement the network, to select the ecosystems, habitats, species and landscapes of European importance, to identify the core areas, corridors and buffer zones, to take action, to improve and manage the network and to review the status of the network

A strategy for assessing the information required, identifying sources of relevant information and reporting on implementation (Harrison 1998) and a project for selecting ecosystems, types of habitat, species and landscapes of European importance and characterising geographical distribution and conservation status of these elements (Nowicki 1998) are to be developed under the work programme for the PEEN.

In some respect, flora and vegetation are the basic determinants of the spatial structure of ecological networks. Readily available standardised information on IPAs will be essential for the development of the Pan-European Ecological Network as far as information on flora and vegetation of European importance is concerned. Both the 'Planta Europa' process and the IPAs initiative should be technically and politically involved in PEBLDS AT1 projects on PEEN information requirements and selection of species and ecosystems of European importance.

Compilation of an ecological network is not only a task and method in the field of nature conservation but also a challenging exercise of geographical information systems' (GIS) application and of spatial analysis. Numerical methods are objective in the sense of repeatability, on the one hand; a GIS approach alone might be too formal, on the other hand, although it could give a more rigid numerical basis for human expert decisions.

The main objective of this paper is to present a national level attempt at designing the Pan-European Ecological Network (PEEN) that is based on Estonian data from a one square kilometre grid (Remm 2000). The other objective is the analysis of the similarity of the suitability map to the existing expert-made ecological network map.

9.2   Map of ecological network as a map of habitat values

Suitable habitat is a key distribution factor for all organisms. The potential and actual distribution of organisms is dependent on the allotment of their habitats or biotopes. Although there are several classification systems for European habitats (CORINE 1991; Davies and Moss 1999; Devillers and Devillers-Terschuren 1996; Påhlsson 1998; Interpretation Manual of European Union Habitats 1999) there is no habitat map covering all of Europe. The CORINE (Co-ordination of Information on the Environment) Biotopes Programme has made a major attempt at habitat mapping. This programme was carried out at sites with significance for nature conservation, regardless of whether the area is formally protected by legislation or not (Möller 1998). Hence, the CORINE biotopes programme has not targeted the entire territory of Europe. Relevant European maps have been worked out that classify land cover, forest, biogeographical or climatic regions, but not habitats. Some

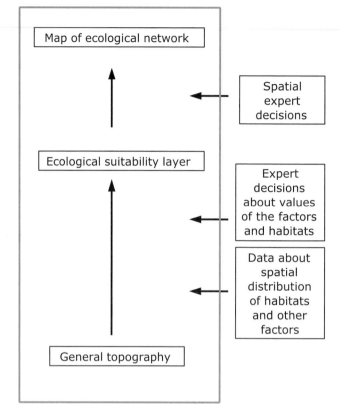

FIGURE 9.1.
Principal layers of a network map.

habitat classifications as potential systems of mapping units are very detailed; for example, the EUNIS (European Nature Information System) habitat classification system contains more than 1500 classification units. Creation of large-scale detailed full-cover habitat maps for larger regions has been too laborious until now.

A map of habitat distribution should be among the main data sources for ecological network mapping, as was stated above, but not the only one. In our opinion a map of the ecological network can be seen as comprising *three principal layers* (Figure 9.1):

- general topographical features such as the coastline, the water network, major roads and place names for locating the depicted network
- a layer of suitability of territorial units for the ecological network
- an ecological network as a decision on nature conservation and territorial planning

Traditional components of an ecological network (core areas, corridors, buffer zones, restoration areas) belong in the third layer. The second layer serves as a supporting tool for decision-making during the compilation of the third layer. Compiling the first and third principal layers of a network map is usually not a serious problem. In most cases until now, maps of an ecological network consisted only of a certain amount of information from a topographical map and expert decisions about the location of ecological network units (Mander *et al.* 1995; Burkhardt *et al.* 1996; Sabo *et al.* 1996). The location of the components of an ecological network can be decided by experts or can also be calculated following a predefined algorithm. The suitability analysis results in a suitability value for every territorial unit of a study area. The critical threshold level of PEEN-suitability determines the relative number of core areas selected. GIS techniques for finding optimum route and spatial discontinuities can be useful in delimiting corridors.

The inclusion of a suitability layer into a network map is not widely accepted, but not a new idea. It is rooted in an ecological evaluation for comparison of natural areas and to facilitate decision-making in conservation. Attempts to set expert decisions into an algorithm of rigid calculation of ecological suitability have been made, for example by Wright (1977), Baschak and Brown (1994), Linehan *et al.* (1995), Roy and Tomar (2000) and van Langevelde *et al.* (2000). These studies either investigate only small or spatially separated valuable territories. In the Estonian case study the suitability analysis according to the idea of the PEEN project has been used. The analysis has not been implemented at the European scale because the data preparation of the PEEN project was still in progress during this study.

There are different ways of working out the ecological suitability of the study area. The direct way tries to estimate the value of territorial units or classes for the ecological network. Woodlands, fields, marshes, etc. are directly evaluated according to predefined methodology. The other approach is the use of indicators. Indicators are usually species and the indicator parameter is the presence, absence or abundance of a particular indicator species. In other words, the indicator approach uses distribution data of species to estimate the ecological value of spatial units. However, indicator parameters are not only species data. Density of contour lines and shorelines of water-bodies are also used as indicator parameters in our case (see Table 9.3).

Several authors have found that there is good reason to be cautious of claims that the spatial configuration and pattern of the remaining habitat can compensate for the overall loss of habitat area (Harrison and Bruna 1999; Trzcinski *et al.* 1999). Therefore, the value of habitats for the ecological network is weighted by the amount of particular habitat, and configuration of habitat patch is not considered directly in this study (see Table 9.2). Habitat configuration is indirectly included by including parameters such as density of borders of land cover units, density of watercourses, density of roads and density of contour lines (see Table 9.3).

All available parameters of landscape and the species distribution have to be evaluated for the suitability analysis of spatial units. Spatial units can be different; these can form a regular grid or be any administrative, management, or ownership units. Numerical expert estimations (PEEN values) are given to the parameters by using existing data and knowledge about the naturalness of habitats, rarity of habitats and species, and relationships between the landscape parameters and biodiversity.

If the evaluation of ecological factors is used on geographically extended regions, such as the whole of Europe, these should be related to biogeographical zones. The PEEN value of broad-leaved forests in the boreal zone is probably different from their value in the Atlantic region. If the ecological importance of the same habitat class is different in different regions, then PEEN values can be altered according to the region, relying on expert decisions or investigations of the importance of one or another biotope for the wildlife.

The importance of PEEN values can also be seen in arranging our knowledge and opinions into a quantitative numerical form. It can be viewed as a common quantitative scale for different data.

9.3   A habitat key for the PEEN project

A simple habitat classification that can be related to available data is needed. Therefore, relatively wide habitat classes are proposed for the PEEN project. These classes can be considered as habitats mainly for large organisms; for smaller organisms these are actually habitat complexes. In combination with biogeographical regions these habitat classes can be seen as potential habitats for particular species.

The creation of map coverage of PEEN habitats is suggested mainly as a modification of land cover data. Most of western and central Europe is covered by the CORINE land cover dataset. This dataset is considered as the main data source. In other regions mainly the use of PELCOM (Pan-European Land Cover Monitoring) project data is proposed (Mücher 2000). Soil data, elevation and biogeographic regions have to be used for delimitation of some habitat classes. The US Geological Survey GTOPO30 Digital Elevation Model is suggested as the main source of elevation data, the Food and Agriculture Organisation of the United Nations (FAO) soil map of the world for soil data, and the Nature/Land Cover information package (NATLAN) of the European Environment Agency or the Digitised Map of Ecological Regions in Europe (DMEER) of the European Topic Centre on Nature Conservation for biogeographical regions. Additional more detailed information could be used if available.

Different generalisation, reliability and level of detail of data sources are important problems in compiling a PEEN habitats map. CORINE land cover data covers only the western part of the study region. The much less detailed PELCOM data does not, unfortunately, cover all the area either. International Geosphere–Biosphere Programme Data and Information System (IGBP-DIS) land cover data is suggested for the European part of Russia not covered by PELCOM data. Biotope maps of CORINE or national mapping projects could be valuable in studies of smaller areas. For a pan-European project these data sources are not among primary data, although they are still valuable for checking and refining the PEEN habitats map.

Classification and a set of instructions (a key) for joining land cover information from different data sources, elevation, and soil data have been worked out for the PEEN project. The first step in the PEEN design concerns Eastern and Central Europe. The draft key of proposed PEEN habitats for this study area consists of 37 units. Additional habitats (e.g. blanket bogs and some Mediterranean classes) can be added if the study area is extended. Much more detailed habitat maps can be generated at a local level in regions where more detailed data sources are obtainable. An example of an implementation of a PEEN habitats key at national level is presented in section 9.4.

9.4   Groups of habitats

The PEEN habitats are classified into six groups according to their naturalness.

The group *artificial* includes man-made surfaces and constructions. These areas do not support wildlife and have minimal primary production. Moreover, these are usually disturbing wildlife and the natural environment in their surroundings. This group contains PEEN habitats: constructions, peat extraction sites.

The group *artificial supporting some wildlife* includes man-made biotopes like orchards, plantations, fields, which are habitats only for limited number of species, most of which are considered pests. Biomass production of these areas is usually high. Suburban areas with gardens and urban greenery support biological diversity. The feeding of birds by humans in suburbs is vital for the survival of many bird species in winter. The group contains PEEN habitats: suburban areas, fields, plantations, and complex cultivation.

Mainly semi-natural habitats have been created by human activities or are maintained in a stable state by some kind of anthropogenic influence. It contains PEEN habitats: wooded agricultural land, scrubs, and Mediterranean scrubs.

Mainly natural habitats stay relatively stable without human influence. Contains PEEN habitats: other grassland, moist grassland, alpine grassland, broad-leaved and mixed forest on peat, other broad-leaved and mixed forest,

Table 9.1. *PEEN habitats containing EU priority habitats*

PEEN habitat	Examples of EU priority habitats
Calcareous grasslands	Nordic alvars and Precambrian calcareous flat-rocks
Boreal coastal meadows	Boreal Baltic coastal meadows
Wooded fens and bogs	Bog woodland
Coastal dunes	Fixed coastal dunes with herbaceous vegetation, decalcified fixed dunes with *Empetrum nigrum*, Atlantic decalcified fixed dunes, coastal dunes with *Juniperus*, wooded dunes with *Pinus pinea*
Inland dunes	Pannonic inland dunes
Fens	Calcareous fens with *Cladium mariscus* and *Caricion davallianae*
Bogs	Active raised bogs, Aapa mires
Salt marshes	Pannonic salt steppes and salt marshes
Mountain coniferous forests	Endemic forests with *Juniperus,* (Sub-) Mediterranean pine forests with endemic black pines

coniferous forest on peat, other coniferous forest, mountain mixed forests, heath, tundra, mountain tundra, alpine scrub, rock, sparse vegetation, glacier, littoral marsh.

PEEN habitats are included in the *priority natural* group if they include a considerable amount of EU priority habitats. Examples are shown in Table 9.1. This group contains PEEN habitat classes: calcareous grasslands, boreal coastal meadows, wooded fens and bogs, coastal dunes, inland dunes, fens, bogs, salt marshes, mountain coniferous forests. PEEN habitats heath and scrub include several EU priority habitats (e.g. Macaronesian heaths and matorral) but none of these is present in the study region of the first stage of the PEEN project. Therefore heath and scrub are included in the previous classes. The PEEN habitat class other grasslands contains EU priority habitats (e.g. Pannonic sand steppes) but the area of priority of other grasslands is considerable only regionally. The very diverse class of other grasslands and several forest classes need further classification if spatial source data for such classification is available.

Water habitats are considered as a separate group because of the aquatic environment, species composition mostly different from terrestrial habitats, and differences in methodological approach in PEEN map compilation. It contains PEEN habitats: inland water bodies, littoral zones, and estuaries.

9.5   Map of ecological network values of Estonian square kilometres

9.5.1 Source data

The following GIS vector data sources have been used for modifying Estonian CORINE land cover data (Figure 9.2):

- map of landscape complexes derived mainly from 1:100 000 soil map
- 1:50 000 digital base map
- boundaries of designated areas on 1 January 2000

Overlay operations and measurements of areas were conformed in ArcView; density of linear features and contours in a square of 1 km × 1 km was measured using MicroStation 95. All the data are stored in a relational database of 1 km × 1 km grid, later also called Estonian square kilometres (Remm 2000). The database of Estonian square kilometres contains data on different geographical parameters for each 1 km × 1 km square that covers at least some part of Estonian land. Midpoints of the database squares are the full kilometre *x* and

FIGURE 9.2.
Source data for calculating ecological suitability layer of the map of Estonian ecological network. VBA, Microsoft Visual Basic for Applications.

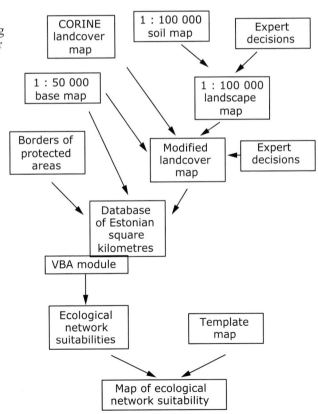

y coordinates of the Estonian 1:10000 base map. Calculations of PEEN values were programmed in Visual Basic (VBA) (Figure 9.2).

The vector format CORINE land cover map of Estonia contains 35 land cover classes and 35 769 polygons (Meiner 1999). Among the 35 classes six are additional classes, not present in CORINE standard classification, e.g. peat extracting areas and transitional woodland/scrub on mire, but used in this paper. This number of classes is too big and not suitable to be used for the development of a habitat map. The following modifications to the Estonian CORINE land cover map were made in order to create a PEEN habitat map:

- most classes of artificial areas (continuous urban fabric, industrial and commercial units, road and rail networks and associated land, port areas, airports, mineral extraction sites, dump sites, construction sites) are joined into the class constructions
- the number of classes of suburban and agricultural units is reduced by joining the classes discontinuous urban fabric, green urban areas, sport and leisure facilities into the class suburban
- the classes pastures, natural grassland and heath are reclassified to derive the classes moist grassland, calcareous grassland, and boreal coastal meadows. The following criteria were used: calcareous grassland occurs on thin soils on calcareous rock, boreal coastal meadow can occur in the coastal zone of the Baltic Sea (less than 1 km from the sea or from littoral marsh or from coastal salt marsh at an elevation of less than 5 m), moist grasslands occur on gley or peat soils, mainly in river valleys. Other grasslands are left in the complex class other grassland. This cannot be further divided until the digitising of the large-scale soil map of Estonia has been completed
- forests on peat soil and mineral soil are separated into distinct classes
- inland water bodies and watercourses are joined into one class: inland water bodies
- dunes are divided into coastal dunes and inland dunes. The latter is represented in only one location in Estonia, on the northern coast of Lake Peipsi

9.5.2   PEEN values and PEEN suitability

Every square kilometre is supposed to have a certain suitability for an ecological network (PS). The suitability of a square kilometre is determined mainly by the habitat structure of the square but also by the location of the square relative to main migration routes, by the land management, and by its legislative position. The direction and magnitude of the influence of these factors on the PS is called the PEEN value (PV) in this chapter. The PEEN value of factors is difficult to measure directly. The only way at the moment is an expert estimation of PEEN values of environmental factors, species occurrence and also anthropogenic influence. PEEN values have been assigned to the habitat classes as non-negative real numbers (0, presence of the factor is excluding the square from the ecological network; 1, neutral influence; 2, twice as good as the average, i.e. the factor is doubling the PEEN-suitability estimation of a square; 10, the factor is multiplying the PEEN suitability of a square ten times) (Tables 9.2, 9.3, 9.4). A multiplicative scale is suggested because it allows the use of the zero value to mark absolutely unsuitable conditions. It means that if a square kilometre contains any amount of absolutely unsuitable habitat, the entire square is considered absolutely unsuitable. The use of absolute unsuitability depends on the scale of spatial units. Zero suitability was actually not applied in this case study.

The influence of one environmental or social factor is determining a partial suitability (ps_i) of a square. The suitability of a square can be conceived as a combination of the influences of several factors. Therefore the overall suitability of a square is a combination (sum, average, or some other) of partial suitability values.

The groups of habitat classes are followed in giving PEEN values to habitats in the following way.

artificial	PV $<$ 0.1
artificial supporting some wildlife	PV $=$ 0.1–0.5
mainly semi-natural	PV $=$ 0.5–1
mainly natural	PV $=$ 1–2
priority natural	PV $=$ 2–10
water	PV $=$ 1–2

The final PEEN value of a habitat class inside a group was given as an expert decision considering the importance of certain habitats for wildlife diversity in Estonia and the distribution of endangered taxa in habitats according to the Red Data Book of Estonia (Lilleleht 1998).

The map of PEEN habitats in Estonia is a generalisation representing a major habitat complex at a site or in a particular square kilometre. The actual distribution of habitats is much more diverse. The dominant habitat usually contains minor patches of smaller habitats. The internal diversity of large habitat complexes could be indirectly estimated by the following characteristics:

- the number of different habitats per unit area. High diversity of larger spatial categories usually coincides with higher diversity of smaller spatial units

Table 9.2. *Ecological network values of land cover classes*

Habitat	Value [PV_i][a]
Constructions	0.01
Peat extraction sites	0.05
Suburban areas	0.1
Fields	0.2
Complex cultivation	0.3
Plantations	0.4
Wooded agricultural land	0.6
Scrubs	0.9
Sparse vegetation	1
Inland water bodies	1.2
Coniferous forest on peat	1.3
Heath	1.4
Other grasslands	1.5
Broad-leaved and mixed forests on peat	1.8
Coastal lagoons	1.9
Other coniferous forests	2
Salt marshes	2.2
Bogs	2.4
Moist grasslands	2.6
Other broad-leaved and mixed forests	3
Wooded fens and bogs	3.4
Littoral marshes	3.6
Fens	4
Coastal dunes	4.5
Inland dunes	5
Boreal coastal meadows	6

[a] In the case of 100% coverage. In other cases the value is calculated according to Equation 9.1.

- density of borderlines of land cover classes
- density of watercourses and coastlines is another estimate of smaller biotopes' diversity not evident from small or medium scale land cover maps
- regions of complex terrain topography usually contain smaller biotopes, as could be seen from the land cover maps, and more of these than in plain regions. Therefore, density of contour lines or any other characteristic of terrain complexity can be used for estimation of biotope diversity not visible on small-scale maps

Table 9.3. *Ecological network values of the density of linear features*

Linear features	Value $[PV_i]^a$
Land cover borders	1.5
Sea coast	1.6
Coastline of inland water bodies (incl. rivers wider than 100 m)	1.4
Watercourses	1.2
Contour lines [recalculated to $\Delta h = 20$ m]	1.3
Roads	0.5

^{a} In the case of unit density $[1 \text{ km km}^{-2}]$. In other cases the value is calculated according to Equation 9.2.

Table 9.4. *Ecological network values of administratively protected areas*

Designated areas	Value $[PV_i]^a$
Ramsar area	1.4
Protected park	1.2
International bird area	1.4
Limited management zone	1.5
Special management zone	2.2
Unmanaged zone	1.1
Strict nature reserve	3
Other	1.1

^{a} In the case of 100% coverage. In other cases the value is calculated according to Equation 9.1.

In addition to the value of habitat categories weighted by their amount, the densities of some linear features as indicators of habitat complexity and the protection category of designated areas weighted by their amount (Table 9.4) are used in calculating the suitability of a site for the ecological network. Designated areas are considered to give additional value to a habitat by giving higher expectation of habitat retention. No conservation arrangement can absolutely guarantee the survival of a particular habitat in a particular site; conservation gives only some additional insurance. The influence of population density or other socio-economic factors could be included into calculations of ecological suitability or into expert decisions about the network.

Partial suitability [ps_i] of areal landscape features has been calculated as

$$ps_i = (PV_i)^{A_i},\qquad(9.1)$$

and of linear features as

$$ps_i = (PV_i)^{D_i},\qquad(9.2)$$

in which A_i is the relative coverage (from 0 to 1), D_i the density relative to unit density (from 0 to ∞) and i the index of the factor influencing ecological network suitability of a square kilometre.

Overall suitability (PS) of a square kilometre or any other complex areal unit is calculated as a log product of suitability values of all categories:

$$PS = \log\left(\prod_i ps_i\right).\qquad(9.3)$$

A square has to be absolutely excluded from the ecological network if the PEEN value of any of the influencing factors equals zero. Overall suitability was calculated in two ways: including the influence of designated areas, and without it. PS is called compound suitability in the first case, and natural suitability in the latter case.

A neighbourhood correction was added to the suitability calculated independently for all squares. Suitability of a square for the ecological network does not depend only on features of the same square; spatial continuity of the network is one of the main goals of ecological networks and therefore the suitability of isolated squares or pixels for the ecological network is small. The value of pixels surrounded by highly suitable pixels should be raised. First the average values of PS have been calculated of eight neighbouring squares for every square. Then the average has been calculated of the PS value of a particular square and the average of the neighbourhood.

The mean PS value of square kilometres containing at least a part of Estonian territory (45 340 km^2) is 0.897, median 1.006; the minimum value is -3.648 and the maximum value is 3.75. The most common network suitability is between 1.0 and 1.5 (Figure 9.3).

The ecological network suitability of protected areas is higher than that of non-protected areas as a rule. The mean natural PS value of square kilometres that contain more than 80% of protected area is 1.34, and the mean natural PS of those square kilometres that do not include protected area is 0.819. The relative amount of protected area correlates positively with the natural suitability for the ecological network (Figure 9.4). Nearly half (47.4%) of ecologically highly valuable areas are under nature protection in Estonia. On the other hand, it means that more than half is not administratively protected.

FIGURE 9.3.
Distribution of natural
suitability values for
the ecological network.

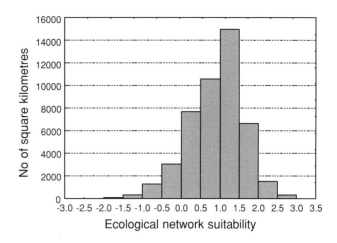

FIGURE 9.4.
Relative amount of protected
areas relative to their natural
suitability for the ecological
network in Estonia.

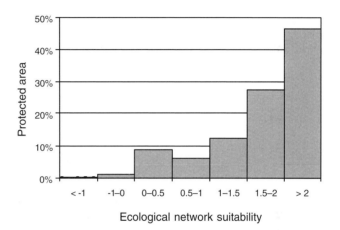

The square with the highest natural PS value is at Laidevahe Bay on the southeastern coast of Saaremaa Island. This region belongs to the West-Estonian Archipelago Biosphere Reserve, which is not counted as a protected area in this analysis. The map of ecological network suitability values indicates several other regions that are valuable for the ecological network but do not belong to the designated areas (Figures 9.5, 9.6, 9.7).

9.6   The development of an expert-designed ecological network in Estonia

Estonia is among the first countries in Europe where the concept of an ecological network has been developed and at least partly implemented. A concept of functional zoning was applied at the national level in the early 1970s, conducted by the spatial planning institutions (Raik 1970). The earlier

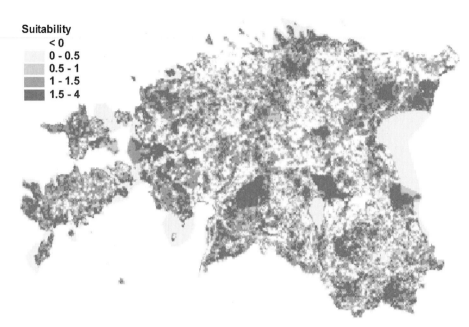

FIGURE 9.5.
Compound suitability for the ecological network in Estonia.

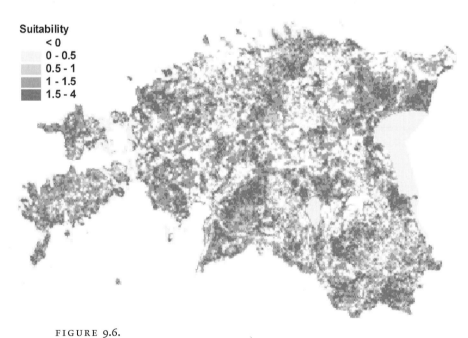

FIGURE 9.6.
Natural suitability for the ecological network. Nature conservation areas are not included in the network suitability calculations.

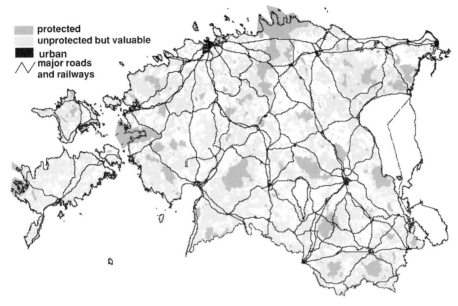

protected
unprotected but valuable
urban
major roads
and railways

FIGURE 9.7.
Protected areas and areas not protected but valuable for the ecological network.

concept, influenced by Rodoman's 'polarised landscape' idea (Rodoman 1974), included functional zoning of the landscape elements into natural zones with the antagonised poles of intensive land use. Emergence of the idea can be considered as a logical consequence of the development of knowledge available, as well as task complexity within planning. Therefore, the planning principles derived from this concept require a strict delimitation of different zones: natural, nature restoration, and recreation zones united into one coherent group, which is proposed to balance the zones selected for agriculture, industry and urban development. In 1979–81 the Department of Geography, University of Tartu, developed a national scheme for conservation and sustainable use of natural resources. For this purpose a methodology for designing a network of compensating areas (an Estonian synonym for an ecological network) was developed (Jagomägi 1983; Jagomägi et al. 1985; Mander 1983; Mander et al. 1988).

The network of ecologically compensating areas was first designed on a scale of 1:200 000. In 1983–88 more detailed maps of the ecological network (1:100 000) were compiled for selected regions in Estonia (North-East Estonia, Tallinn region, Hiiumaa, Saaremaa, North-West Estonia). These maps were intended to be used as spatial reference information for development programmes until 2005. At the same time, in the early 1980s, a methodology for ecological networks at the micro-scale level, consisting of buffer zones for protected wetlands (Mander 1980), roadsides (Mander 1983) and water bodies (Mander 1995), was worked out.

Despite a wide range of research and implementation capabilities in the country the ecological network concept was not implemented in environmental protection practice until the mid-1990s. Only the Act on the Protection of Coastal Areas (1994) and the Act on the Protection of Surface Waters (1994), prohibiting all building activities within 200 m of the coastline and lake shores, provided the first pieces of legal background for designing an ecological network as a part of county spatial planning processes. However, at present the ecological network concept is already better reflected in strategic planning documents. The need for formation and improvement of the network of protected areas can be found in the Estonian Environmental Strategy (Estonian Ministry of the Environment 1997). It provides 'to establish /by the year 2010/ a network of nature reserves corresponding to EU recommendations where zones of strict protection have to cover up to 5% of the terrestrial area of Estonia'. The Estonian Environmental Action Plan (Estonian Ministry of the Environment 1998) sets for the period 1998–2000 such actions as: (a) setting principles for ecological network design, (b) defining elements of the ecological network and (c) for the period 2001–06 to update and develop the ecological network concept at the national and regional level.

In 1999 the Governmental decree for the second phase of county planning (1999–2002) on 'Defining environmental conditions for the development of land-use and settlement structure' was issued. The county spatial plan is considered to be a main development plan, which sets the legal framework for land use and other activities. The main tasks of the second phase of county planning include:

- design of green network at county level (planners are using the term 'green network' instead of 'ecological network')
- defining valuable cultural–historical landscapes

Consequently, by 2001 each of the 15 counties must prepare a map of the ecological network as one of the layers of the spatial plan. For that reason, at least two methodologies will need to be elaborated: one for designing green networks and the other one for defining valuable cultural landscapes.

Last but not least, the long-term strategy *Estonia – Vision 2010* (Jagomägi *et al.* 2000) contains the schematic map of the Estonian green network (Fig. 9.8). This is the first GIS-based attempt at ecological networks in Estonia; however, it is a typical expert-designed system without using a strong algorithm. Here, the application of at least two criteria – area (shows the overall compensatory capacity) and protective value (priority of habitats or protected species at the international level) – is thought necessary.

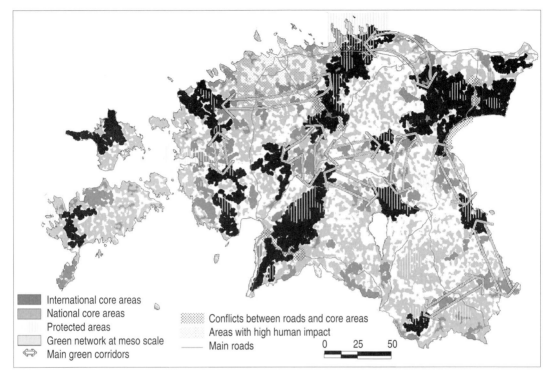

International core areas
National core areas
Protected areas
Green network at meso scale
Main green corridors

Conflicts between roads and core areas
Areas with high human impact
Main roads   0   25   50

FIGURE 9.8.
Estonian 'green network' according to *Estonia – Vision 2010* (Estonian Ministry of the Environment 2000).

The ecological network of *Estonia – Vision 2010* includes forests, wetlands, natural and semi-natural grasslands, lakes and seashores to a depth of 6 m. Depending on their location and their utilisation rate, arable land and the rural areas on the outskirts of towns, as well as parks, garden suburbs and residential areas in town, may be partly included in the network. Protected areas that are independent and have been determined by law are regarded as the core areas in the ecological network. In the first round, the present plan stipulates the inclusion of natural zones with an area of more than 100 km² as core areas of international importance. After the cartographic generalisation such large expanses form 12 comparatively compact core areas in Estonia (Jagomägi *et al*. 2000). This is similar to the earlier but less detailed schemes of an ecological network (Mander *et al*. 1995).

The maps of suitability of Estonian square kilometres for the ecological network (Figures 9.5 and 9.6) show a great similarity with the map of ecological networks of *Estonia – Vision 2010*. On one hand, this is influenced by the equal distribution of ecologically important areas, but, on another, it demonstrates a quite good expert estimation. However, in the further elaboration of planning strategy, it is necessary to specify additional essential criteria such as endangered and flagship species, habitat for selecting elements of regional importance and defining the structure of ecological network at the lower level. The aim of the strategy at the lower level is to guarantee the maintenance of core areas of national importance in their present area, sufficient migration of species through the corridors, and enhanced control of material fluxes through the landscape.

9.7   Conclusions

A methodology has been designed for suitability analysis of territories for ecological networks as a decision support system. An expert-estimation-based habitat classification (a land cover classification improved by soil and landscape data) is proposed as a basis for creating the suitability maps. A great similarity was found between the suitability map for the ecological network and the existing expert-made map of the ecological network presented in the work on *Estonia – Vision 2010*. However, in the case of less similarity, creation of a suitability layer for ecological network maps might help decision makers. Likewise, the evaluation of factors and the creation of a suitability layer might make spatial decisions more scientifically justified. Therefore, we propose the present methodology for designing the Pan-European Ecological Network at the national level.

The application of suitability criteria for ecological networks design helps to reveal, evaluate and exploit the impact of protected areas and sparsely populated territories on the environment in the broader sense.

10

Ecological 'black spots' within the ecological network: an improved design for rural road network amelioration

10.1 Introduction

The presence of a well-developed road network in a region is a *sine qua non* for both economic development and efficient use of land resources (OECD 1986; Macpherson 1993; Tolley and Turton 1995; Jaarsma 1997). Rural road networks are planned to facilitate human movements.

Where there is fragmentation of habitat of both plant and animal populations, there is a need to facilitate movement of wildlife. For this purpose, ecological networks are planned. These networks are composed of core areas, buffer zones, and connecting ecological corridors (Jongman 1999). Ecological networks consist of large patches, which are the core areas, with surrounding buffer zones, mutually connected by long stretches of linear elements, which are the corridors (van Bohemen *et al.* 1991). Smith (1993) defines these corridors as 'linear conservation areas' or 'greenways'. The term greenway has been applied to a wide range of landscape planning strategies, concepts, and plans. Nevertheless there is still little agreement on terminology (Ahern 1995). For this reason, within the context of this chapter, we use the term ecological network, according to Jongman (1999) and van Bohemen *et al.* (1991).

Well-known conflicts arise where the ecological networks and greenways meet human highways, where nature conservation conflicts with roads and the traffic they carry. Forman and Alexander (1998) and Spellerberg (1998) present surveys of the extensive literature in this field. Much of the research refers to the conflicts in the core areas and/or the conflicts with motorways. This chapter also addresses conflicts that emerge when ecological corridors cross minor roads. The ecological 'black spots' that can appear when the ecological network crosses the rural road network are especially problematic for the wildlife. It is at these points that the function of the ecological network can be most disturbed by barrier effects, resistance to movements, and wildlife colli-sions, all due to motorised traffic.

In contrast with the core areas, the prevailing land use in the corridors is gen-erally not nature conservation. A well-functioning road network is essential to facilitate neighbouring land uses in or near the corridor zone. The challenge for planning is to serve both human and wildlife movements in a well-balanced way. It was with this in mind that Jaarsma (1997, 2000) advised traffic and transportation planning on a regional scale. The results of the advice are re-flected in this paper, which elaborates the possibilities to avoid or at least reduce the conflicts between the ecological network and the road network. Section 10.4 discusses the concept in depth, and three Dutch case studies are discussed in section 10.5. Section 10.2 elaborates the problem of roads and their traffic in relation to wildlife. A subdivision of the road network is explained in greater detail in section 10.3.

10.2   The relations between wildlife, roads and their traffic flows

Four adverse effects of roads on wildlife species can be considered (after van der Fluit *et al.* 1990; van Bohemen 1995):

- destruction or alteration of habitat due to construction
- disturbance of habitat along the roads due to noise, vibrations, car visibility, artificial lighting, etc.
- physical barriers created by roads that can vary from increased resistance for movement to separation of functional areas
- collisions during crossing that result in injury or death of wildlife due to motorised traffic

The first two effects directly influence the amount of wildlife habitat and its quality. Habitat loss due to road construction is a single event and in general relatively small. It depends, for the most part, on the construction width of the road. However, where the disturbance leads to road avoidance (Forman and Alexander 1998), ecological impacts are both larger in scale and continual in occurrence. As far as the disturbance of habitat is related to the noise load (see, for example, Reijnen *et al.* 1995), the range of the disturbance and thus the amount of habitat area with loss of quality are related to both traffic volumes and speeds. These two factors encroach upon wildlife movements alongside the road.

The barrier effects of the road and its traffic encroach upon those wildlife movements that traverse the road. They affect populations of insects (see, for example, Vermeulen 1994), reptiles and amphibians (see, for example, Vos and Chardon 1994), breeding birds (see, for example, Reijnen *et al.* 1995), and

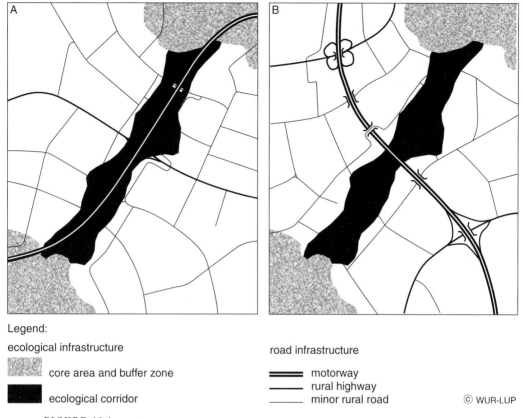

Legend:

ecological infrastructure

    core area and buffer zone

    ecological corridor

road infrastructure

══════ motorway
───── rural highway
───── minor rural road

© WUR-LUP

FIGURE 10.1.
Schematic representation of two different orientations between the ecological
network and the major road links: (A) parallel and (B) transverse.

mammals (see, for example, Mader 1984; Lankester *et al.* 1991). Owing to the
barrier effects of roads and traffic, the species will decide not to cross the road at
all, follow another route, or risk a collision (van Langevelde and Jaarsma 1997).
Roadkills are a major source of mortality, but they rarely actually limit the
population size of a species. The barrier effects can, however, subdivide populations, which results in both demographic and, in all probability, genetic consequences (Forman and Alexander 1998).

    The adverse effects of roads on species are related to the direction of the move-
ment of the animals. The configuration of the connection between the ecological network and the road network is very important for the impact of these
effects. Figure 10.1 presents a schema for two different orientations between
the ecological network and the major road links: parallel (A) and transverse (B).
The road network in the figure is assumed to be hierarchical, with a dominant
direction for traffic flows, despite the presence of minor road links in the network that run perpendicular to this dominant direction.

When the ecological network and the major rural roads in a region run parallel to each other, the interactions between both networks primarily result in disturbance alongside the road. By contrast, when the ecological network crosses the rural road network, the barrier effects resulting from this interaction have far more relevance and extend beyond the diminishing quality of the habitat area. In this situation, the mortality effect of roads, i.e. the number of individuals killed by traffic, and the movement of individual species are far more important issues (Verboom 1994). The same is true for those core areas intersected by minor roads.

There are complex relationships between the barrier effect and the risk of collision during crossing, which is the risk factor on one hand, and the number of traffic kills during dispersal, the dispersal death rate, and the number of successful crossings of individual species on the other hand, which is the crossover rate (Verboom 1994). To quantify these effects, relationships between traffic and road characteristics must be found.

For instance, a wider road encourages both higher traffic volumes and higher speeds. This, in turn, reduces the chance of a successful road crossing (as formulated by van Langevelde and Jaarsma 1997), since the intervals between vehicles become much smaller. Moreover, the wider the road, the more time an animal needs to cross the road and the less chance it has to succeed. In addition, an increase of volume may lead to such a flow of vehicles that individuals are prevented from crossing the road. Finally, an increase of volume also determines the noise level (see, for example, Reijnen *et al*. 1995).

Despite the fact that the foundations and the derived formulae for the relationships between both the noise level and the chance of collision with the traffic volume are quite different, there is an important comparison to be made. High traffic volumes cause high noise levels and simultaneously increase the chances of collision. This works the other way around as well, because a decrease to a certain volume will also diminish both the noise level and the probability of collision. However, such a decrease in the noise level and the chance of collision on a minor road far exceed the increase of both these effects caused by the increase of traffic volume on a major road[1]. Concentration of diffuse traffic flows in rural areas on a few major roads is thus a useful measure on a regional scale to decrease both disturbances by noise and the chance of collision during crossing. This idea is expanded in the next section with a categorisation of roads within the road network.

[1] This is because of the exponential character of the relationships with traffic volumes. For example, a reduction of the volume from 1000 to 500 cars per day will reduce the noise level from 68 to 65 dB(A). An increase from 2000 to 2500 cars means an increase of only 1 dB(A), from 71 to 72 dB(A). When we 'standardise' the collision probability in the first situation (1000 cars per day) for a certain road and species as 100, this probability is approximately 50 for 500 cars, 200 for 2000 cars and probability 248 for 2500 cars per day.

10.3   The road network dismantled

The rural road network may be subdivided into three groups:

- motorways for fast long-distance trips
- rural highways for regional access and regional trips
- local access roads for local access only; the terms 'low-volume road' and 'minor rural road' are also in use

Despite this hierarchical subdivision, we stress that the three groups function as one system. Therefore, regional traffic and transportation planning must be carried out for the rural road network as a whole (Jaarsma 2000).

All groups of roads have their ecological effects. Current research in this field, however, focuses particularly on motorways (Reijnen *et al.* 1995; van Bohemen 1995; van Bohemen *et al.* 1991; van der Fluit *et al.* 1990). This can easily be explained by the larger zones of destroyed habitat and decreasing habitat quality along these major roads. Not motorways, but rural highways and local access roads, minor roads, determine the rural road network (OECD 1986). The paved roadway width and the verges of minor roads are smaller, often much smaller than those of major roads. The more urgent impact of minor roads on habitat fragmentation is their presence in rural areas. In the Netherlands, their average mesh size is only 1.5 km, compared with about 30 km for motorways. In industrialised countries with a high density of population, the minor roads, especially, often absorb high volumes of traffic. Here, traffic problems result in (Jaarsma 1997):

- increasing costs of road management due to damage to the verges and the construction, caused by overloading
- lack of traffic safety, the number of personal injury accidents per vehicle kilometre on local access roads is ten times of the number on motorways
- declining quality of life for residents owing to annoyance, emissions, and reduced safety
- increasing habitat fragmentation for wildlife

Despite the presence of a rural road network with a sufficient mesh size, increasing volume has caused growing and complicated problems. Today, the network even fails to absorb the current volumes of traffic in a safe and efficient way (Jaarsma 1997). To solve or at least mitigate these problems, both major and minor rural roads need to be considered. In addition to the traffic problems, environmental problems including habitat fragmentation must be an important aspect of an integrated approach to the rural road network. The next section explains just such an approach.

10.4   Regional traffic planning as a solution

Van Langevelde and Jaarsma (1997) advocate an integrated regional planning approach to address both traffic and environmental problems. This marks a transition to road planning within a wider perspective, as advocated by OECD (1986). It also marks a transition from planning for road links to planning for road networks (Jaarsma 1997). This approach acknowledges the fact that a planning intervention may affect a larger area than the direct surroundings of its subject. This holds true for both traffic and wildlife. Moreover, problems at one place may be remedied by interventions at another place.

The spatial concept of 'Traffic Calming' can guide planning interventions at a regional scale. This concept is a spatial organisation principle that contains several engineering consequences for infrastructure. A traffic-calmed area is an area containing minor roads with an access function only, enclosed and accessible by major roads.

Traffic calming began in the late 1960s in built-up areas, with speed reduction. During the 1970s this idea became more widely known, with the concept 'woonerf' in the Netherlands, followed by the German 'Verkehrsberuhigung' in the 1980s (Macpherson 1993). The basic principle is one of integrating traffic in residential areas, giving priority to people's needs, not the needs of traffic. In a residential area, traffic is unavoidable, but it should be subsidiary to other spatial functions. Traffic flows are concentrated onto distributor roads, which are capable of coping with these flows (Jaarsma and van Langevelde 1996). The ratio of cars in the area and on the distributor roads is related to the type of technical measures instituted. Figure 10.2 shows three models for urban traffic calming as a broad concept.

The concept of a Traffic-Calmed Rural Area (TCRA) takes these broad ideas on traffic calming from built-up areas and transfers them to the rural area (Jaarsma et al. 1995). The starting point is the desired spatial function of the rural area,

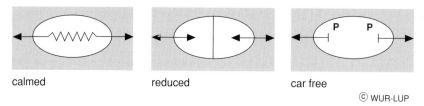

calmed   reduced   car free

© WUR-LUP

FIGURE 10.2.
Schematic representation of three models for urban traffic calming. The model at the left still allows cars in the residential space, but these have drastically reduced speeds. The model to the right does not allow car traffic in the residential area at all. The model in the middle divides the residential area into sections. Car traffic from one section to another is only possible via the distributor road.

not the actual traffic flows. In addition to the needs of people, the needs of nature are also considered. Usually, the residence functions for both inhabitants and leisure-seekers, and for wildlife, will be stressed, rather than the flow function for through traffic. Roads within the region will mainly have an access function, with an accompanying modest technical layout. The region is surrounded and made accessible by rural highways and/or motorways, with a flow function. The underlying idea is a clear separation of space for living and space for traffic flows (Jaarsma and van Langevelde 1996; Jaarsma 1997).

Although this underlying idea is the same for both rural and built-up areas, large discrepancies occur in the desired traffic characteristics and the measures needed (taken) to realise these. In the TCRA, traffic speeds must be modest, from 40 to 60 km h^{-1}. However, it is not desirable for a rural-bound driver to drive for too long on such a low-standard, type B access road. Therefore a second type, with somewhat higher technical standards and speeds from 60 to 80 km h^{-1}, is planned: the type A access road. This type A access road is the only one connected with the network of rural highways. Rural highways, in their turn, form the backbone for the regional rural traffic flows. They also connect the rural area to the motorway network. On a global scale, Figure 10.3 illustrates some options for a categorisation of road types within a traffic calmed rural road network. All models belong to the type presented on the left in Figure 10.2. Traffic is allowed to use all rural roads, but slower speeds on B-type access roads will stimulate use of the routes along major roads. A major

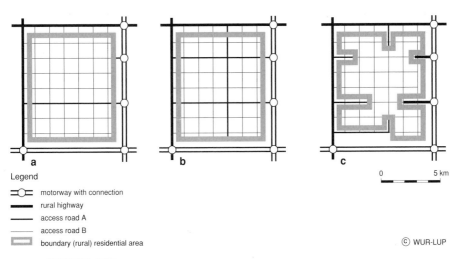

Legend

⊐○⊏ motorway with connection

▬▬▬ rural highway

──── access road A

──── access road B

▭ boundary (rural) residential area

0   5 km

© WUR-LUP

FIGURE 10.3.
Schematic representation of three options for a hierarchical rural road network, based on traffic calming: **a** residential area with two types of access road; **b**, as in **a** but with more type A access roads; **c**, residential area with introversions.

difference between the three options in Figure 10.3 is the distance to be travelled on a B-type access road, before reaching a major road. For this reason the accessibility of destinations within the rural area as well as the level to which the flows can or cannot be concentrated will vary.

The concept of the TCRA decreases traffic speeds on minor rural roads. With a well-designed network this may result in a reorganisation of traffic flows:

- diffuse volumes at the local access road level will be concentrated on a few rural highways
- remaining traffic flows at local access roads are rural-bound either in origin and/or in destination along such a road
- traffic volumes and speeds within the region will decrease

It is expected that such a reorganisation will counter both the lack of traffic safety and habitat fragmentation within the region (Jaarsma and van Langevelde 1996).

The concept of the TCRA provides a method for the (re)assignment of traffic functions and adaptations of the design resulting in a reorganisation of traffic flows: diffuse flows of through traffic in the rural area will be forced to use a few nearby rural highways and, if present, motorways. Traffic volumes and speeds within the traffic-calmed rural area will decrease. The concept leads to a few roads with a traffic function and related higher volumes, whereas most rural roads only have a local access function with low volumes, restricted speed and an optionally adapted design.

From the viewpoint of accessibility, Model **b** is superior to the models **a** and **c** of Figure 10.3. In model **a** traffic flows may be very diffuse. Model **c** seems to offer good opportunities to concentrate flows on the major road network, but trip lengths will be substantially higher than in Model **b**. Therefore, Model **b** seems to be the most realistic one.

10.5   Case studies

This section focuses on the results of three Dutch case studies. The first, in the south-west part of the Netherlands, close to the Belgian border, addresses the multiple land use conflicts that have arisen. The second discusses the problems resulting from a former national road, dividing one core area within the ecological network in the northern part of the Netherlands into two. The third case study illustrates the problems of an extended rural road network in a corridor zone of the Dutch national ecological network. This case is also in the northern part of the Netherlands.

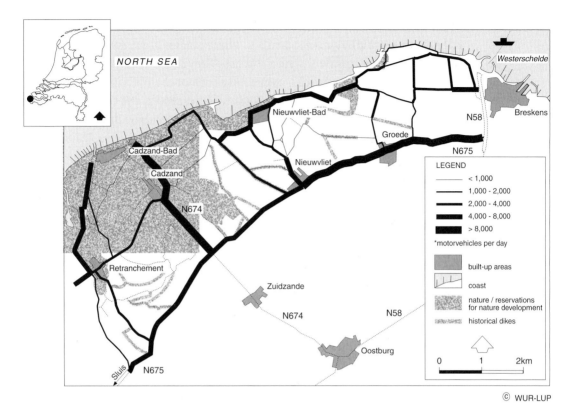

FIGURE 10.4.
Traffic volumes, as calculated for the autonomous development in Zeeuwsch
Vlaanderen (Jaarsma and Baltjes, 1995).

10.5.1 Zeeuwsch Vlaanderen

In the coastal region of the western part of Zeeuwsch Vlaanderen, conflicts arise from the competing land uses of outdoor recreation, nature, and
agriculture. The motorised recreational traffic flows to the beaches have determined the desired road layout. The main access road is a provincial road (N675),
running parallel to the coast, a few kilometres inland. It connects the villages
with the national road (N58) to the ferry harbour near Breskens. However, traffic flows are diffuse, as shown in Figure 10.4. In particular, the appearance of
a considerable flow of through traffic on the minor roads alongside the coastal
dunes frustrates both recreational users and nature development in the long
stretches of coastal zone.

As a solution to this problem, the province of Zeeland has suggested a seg-
mented structure, where the long stretches of coastal zone between the beaches
and the N675 are divided into nine individual compartments that alternate
and result in less crowding. The opening-up is co-ordinated with this principle

by means of a so-called 'slagen' structure, i.e. connections between the provincial major road (N675) and the beaches. These 'slagen' are rural highways, ending in a dead end (cul de sac) at the coast. They are only situated within the more densely populated areas, connecting each of the five villages (Breskens, Groede, Nieuwvliet, Cadzand and Retranchement) with the coast. In the less crowded zones, traffic calming takes a lower priority than the need for traffic. Here, type B access roads are planned. For each crowded zone, Jaarsma and Baltjes (1995) investigated alternative locations for the 'slagen.' They calculated traffic volumes and their impacts for these alternative network solutions[2]. In comparison with the so-called autonomous development in Figure 10.4, a 'slagen' structure decreases the traffic performance within the segment itself. Traffic flows on the provincial road (N675) increase proportionately. Differences between alternative locations of the rural highways in the crowded zones are often considerable. Despite an overall reduction of traffic performance of 6%, several alternatives are under study that include a local increase in performance, by as much as 60%.

In this study the impacts on habitat fragmentation are defined indirectly. Alternative network solutions with low traffic volumes on access roads near areas reserved for nature conservation and/or development are considered to be favourable. Areas such as these are planned at three locations in the coastal area: north of Groede; west of Nieuwvliet-Bad; and between the coast and Retranchement, including a circle of forest around this historical town (Jaarsma and van Langevelde 1997).

This case study shows the advantages of integrated traffic planning in this intensively used region. Among its advantages, it offers the building blocks for both nature development and a well-developed road network.

10.5.2 Gerdyksterwei

The Gerdyksterwei is a former national road, between the villages of Gorredijk and Beetsterzwaag in Friesland, one of the northern provinces of the Netherlands (Figure 10.5). Since it is bypassed by the A7 motorway, the planned traffic function of the Gerdyksterwei is restricted to access road. However, the technical standard is still high. It is therefore not surprising that many drivers between Gorredijk and the nearby town of Drachten prefer this route to the functional route along the A7. Therefore, daily traffic volumes on the Gerdyksterwei, and in the village of Beetsterzwaag, 4100 and 5300 motor vehicles per day, respectively, are high for an access road. The capacity of the nearby

---

[2] To calculate the traffic volumes, first the present volumes are supplemented with the expected growth (autonomous development). Next, these 'autonomous' volumes are assigned to the road network as proposed in the alternative network solutions.

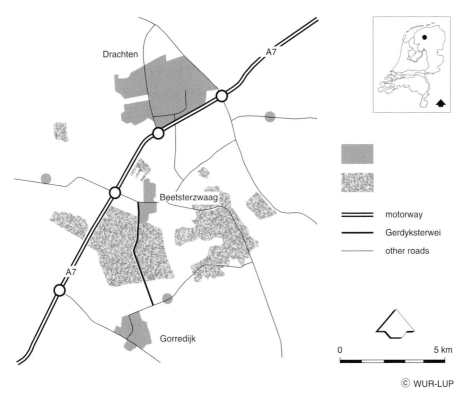

FIGURE 10.5.
The Gerdyksterwei in its regional context (Jaarsma and Van Langevelde 1997).

motorway allows for carrying these vehicles, since it carries only 30 000 vehicles per day.

The Gerdyksterwei crosses an extensive wooded area, which is a core area in the Dutch national ecological network. Owing to the high traffic volumes, the chance of collision when crossing the Gerdyksterwei is considerable for species such as the roe deer *(Capreolus capreolus)*, the pine marten *(Martes martes)*, and the otter *(Lutra lutra)* when it is reintroduced. Here, traffic calming means that priority has to be given to wildlife, not to through traffic.

Van Eupen and van der Veen (1995) compare the probability of collision for crossing in the present situation with autonomous development (an increase of more than 1000 vehicles per day in 2005, based on expected growth), and a planned development of a TCRA, without through traffic (with a decrease of 2400 vehicles per day that would be diverted to the A7). In the latter situation, the lower volumes allow for a reduced road capacity. Carriageway width may then be reduced from 7 m, where it currently stands, to 5 m or even 4 m. With a formula for successful road crossing (van Eupen and van der Veen 1995), the impacts of road mortality were calculated for the roe deer, the pine marten, and the otter. The results of these different calculations showed only small differences. Most traffic victims are hit during the night. Therefore, nightly volumes are estimated on the assumption of a quarter of the daily traffic flow appearing between 7 p.m. and 7 a.m.

In the autonomous development, the mortality effect will increase by approximately 25%. With the lower, TCRA-implemented volumes, the mortality effect is predicted to decrease by 66% compared with the actual situation. This improvement may be made even larger by narrowing the present paved roadway width (7 m): 75% and 79% reductions are calculated for a paved roadway width of 5 m and 4 m, respectively. So the number of roadkills during crossing can be reduced by traffic calming here.

If traffic calming on the Gerdyksterwei is realised by an effective reduction in speed, the travel time between Gorredijk and Beetsterzwaag will increase by approximately 45%. However, with a local closure that forces local traffic onto the A7, this will be 70%. For longer trips, such as from Gorredijk to Drachten, differences between both routes are much smaller. Use of the motorway A7 for that trip takes about 4% more time, although the route is approximately 20% longer (van Eupen and van der Veen 1995). Since the A7 is located in an open landscape, an increasing noise level is a disadvantage. However, this increase is only 0.3 dB.

This case study addresses the opportunities for rural traffic calming, in this case above all favouring wildlife, in combination with better use of an existing motorway.

10.5.3 Ooststellingwerf

This case deals with a more complete application of the TCRA concept to a region with safety problems resulting from 'rat run' traffic (Jaarsma *et al.* 1995; Jaarsma and van Langevelde 1996, 1997).

Ooststellingwerf is a rural municipality in the south-east part of Friesland, a northern province of the Netherlands. The study area (Figure 10.6) is approximately 80 km². It is situated between the small town of Oosterwolde, with 9600 inhabitants, and the villages of Haulerwijk (population of 3200), Waskemeer (population 830) and Donkerbroek (population 1900). Haule has a population of 570 and is a small, linear village in the centre of the study area. The N381, a rural highway west of Oosterwolde and Donkerbroek, connects the study area with the national network of motorways. The southern boundary is formed by the N919, a smaller highway connecting Oosterwolde with the eastern part of the Netherlands. The other major roads in the study area are the N918 between Oosterwolde and Haulerwijk, and the N917, the connection between Haulerwijk and the N381, via Waskemeer. The rural area is mainly

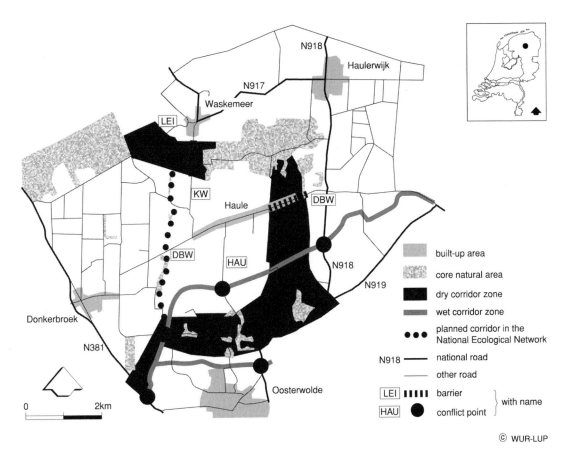

FIGURE 10.6.
The National Ecological Network (NEN) in the Ooststellingwerf region.

grassland, with some small forested zones. The area is designated as an important corridor zone for both wet and dry habitat in the National Ecological Network.

The actual categorisation of the road network is not very clear. Although the technical standard of the rural highways is relatively low, several access roads have a rather high technical standard. As a result, traffic flows in the study area are diffuse. Figure 10.7 shows average annual daily traffic volumes (AADT) for the somewhat autonomous development. This includes independent developments over the next ten years, such as road reconstruction outside the study area, new residential areas, etc. The autonomous development refers to the impacts presented below.

For most roads, the autonomous volumes in ten years' time are expected to be about 11% above the present level. However, such an increase of volume for the entire region is not desirable. Residents already complain about high speeds and rat run traffic. Furthermore, the safety figures are unfavourable

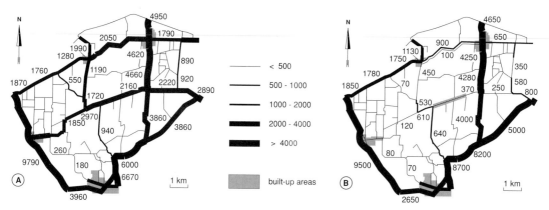

FIGURE 10.7.
Traffic volumes (in motor vehicles per day) in the Ooststellingwerf region:
(A) autonomous and (B) TCRA-implemented development (Jaarsma *et al*. 1995).

(Jaarsma *et al*. 1995). Some roads in the area (DBW, HAU, KW, LEI in figure 10.6) act as barriers, especially for larger animals (van Langevelde and Jaarsma 1997).

The TCRA concept is applied to the rural area in Figure 10.6, and is bounded by the major roads N381, N919, N918 and N917. The proposed speed for the major roads is 80 km h^{-1}, the legal limit. A speed of 60 km h^{-1}, is the starting point for the minor roads. Based on this network, traffic volumes and travel times per road link are calculated with an assignment of flows to the more functional routes. Next, the impacts of the TCRA concept are calculated by looking at accessibility, traffic performance, safety, and habitat fragmentation. These impacts are compared to those resulting from the autonomous development.

Figure 10.7B shows the traffic volumes (AADT) for the TCRA-implemented situation. The differences from the autonomous situation become quite clear. In the autonomous situation, diffuse flows will continue. On nearly all access roads the situation will become worse than it is at present. In the TCRA-implemented situation, the N919 serves the heaviest flows, now crossing the area (from Donkerbroek via Haule eastward and *vice versa*, but also from Waskemeer via Haule eastward and *vice versa*).

The number of daily vehicle kilometres travelled in the TCRA-implemented situation is about 13% lower than in the autonomous development. However, conforming to the idea of the TCRA concept, there is a big difference between major and minor roads. The distance travelled on minor roads in the TCRA-implemented situation is only 43% of that in the autonomous development, whereas the major roads have a small increase of 7% (Jaarsma *et al*. 1995).

The impact of fragmentation of roads in this case study consists mainly of barriers produced by the presence of roads and the resulting traffic. Within that

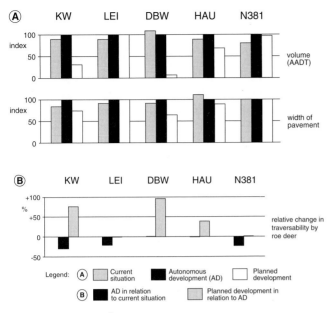

FIGURE 10.8.
Changes in (A) road and traffic characteristics and (B) effects on the roe deer for the actual situation in 1994, the autonomous development in ten years, and the TCRA-implemented situation in the Ooststellingwerf region.

context, the concept of traversability is elaborated. Traversability is the probability of a successful road crossing by a species. For the theoretical probability, an exponential formula is derived using the factors of decisive traffic volume, paved width of the road, and body size and traversing speed of the species. For the roads indicated as bottlenecks in the corridor zones (Figure 10.6), Figure 10.8A shows changing volumes and paved roadway widths. For the same roads, Figure 10.8B indicates changing the traversability for the roe deer (van Langevelde and Jaarsma 1997).

In the autonomous development, all volumes except one (DBW) will grow. Three roads need to be widened to offer the desired capacity. As a result, traversability will decrease by 20–30%. In the TCRA-implemented situation, by contrast, three roads (KW, DBW and HAU) show a considerable volume decrease. The paved width can then be reduced. As a result, traversability will improve by 40–95% in relation to the autonomous development. For the major rural highway, the N381, differences among these three situations are modest.

Introduction of the TCRA concept in this region improves traffic safety as well. Based on a combination of vehicle kilometres travelled and accident rates per type of road, the yearly number of injuries will rise from the current 51 to 54 persons in the autonomous development. In the TCRA-implemented situation the number will decline by 20% to 43 people. A comparable decline, at a much lower absolute level (from 2 to 3 persons), is calculated for the number of traffic deaths (Jaarsma *et al.* 1995).

It is concluded that the concept of TCRA will not substantially diminish the accessibility of the area. For this purpose, Jaarsma *et al.* (1995) quantify accessibility as the part of the road network that is fit to be travelled within several minutes from a built-up area. It must be considered, however, that distances within this TCRA are modest, about 10 km in a straight line at most. Therefore, travel times also appear to be modest. In addition, the scale of the TCRA must not be too large.

10.6   Discussion

This chapter discusses the possibilities to avoid or at least reduce the conflicts between the ecological network and infrastructure. The distinction between major roads, such as motorways and rural highways, and minor roads as local access roads is applied for this context. All groups of roads have their ecological effects. Attention often focuses on major roads, with their larger zones of destroyed habitat and the decreasing habitat quality. However, the role of minor roads is also important because of their frequent presence and intense use in many cases. Their length, in combination with frequent interruptions by driveways to adjacent farms and plots, restrains the use of effective mitigating measures such as fencing in combination with tunnelling.

Most of the effects of roads on species are related to traffic volumes. The overall diminishing effects have been explained, when present diffuse flows on access roads are concentrated as much as possible on major roads. To achieve this, an integrated regional planning approach is advocated to address both traffic problems and environmental problems. The spatial concept of the Traffic-Calmed Rural Area is just such an approach.

The three case studies that are presented illustrate the potentials of this approach. For the ecological network, a considerable improvement in successful road crossings by a species is possible, especially for the minor rural road crossings. One may argue that rural highways and motorways have to absorb the concentrated traffic flows, inevitably increasing the traffic volumes. However, this hardly increases the ecological impacts of these major roads. Furthermore, the present practice of fencing and tunnelling may be continued. The benefits alongside a long stretch of minor roads by far exceed the modest negative aspects alongside a much smaller stretch of major roads. As a result, the ecological networks achieve an overall gain in nature area.

The case studies are all located in the Netherlands. However, the problems described also appear in those regions with a dense network of minor roads, in other industrialised countries. The concept of the TCRA can also be applicable there, provided that the road network is dense enough to allow for alternative routes that are not too long and circuitous.

10.7 Conclusions

Infrastructure and the resulting traffic flows conflict with the ecological network. The role of minor rural roads is thus often underestimated. This network is widespread in the countryside and often used quite intensively. Owing to barrier effects of roads and their traffic, species will either decide not to traverse a road at all, following another route, or will undertake the risk of a collision. In contrast with major roads, mitigating measures such as fencing and tunnelling are not applicable in practice alongside rural roads. Therefore other solutions must be found.

Rural traffic calming can be such a solution. In three Dutch case studies the concentration of diffuse flows was found on a limited number of major roads. These can facilitate the ecological networks, by diminishing disturbance and road kills on a regional scale. These impacts can be further improved by the use of fencing and tunnelling alongside the major roads. For local people, the related decrease of accessibility of the rural area can be limited, if the distances within the traffic calmed area are about 10 km at most as the crow flies. Traffic calming also brings benefits to local people, such as increased traffic safety and improved quality of life within the rural area.

The conclusion is that rural traffic calming is a necessity in densely populated areas with a complex network of intensively used roads. A new design for the rural road network may facilitate the movement of both people and nature and may solve some of the ecological 'black spots' within the ecological network.

RENATO MASSA, LUCIANO BANI, MARCO BAIETTO,
LUCIANA BOTTONI AND EMILIO PADOA SCHIOPPA

11

An ecological network for the Milan region based on focal species

11.1 Introduction

The Milan region is one of the most urbanised areas in southern Europe. In many anthropogenic temperate landscapes, such as the lowland of north Milan, most remaining natural or semi-natural habitat fragments consist of lowland forest. The land is highly fragmented and nature is under severe pressure from intensive land use. Many functions compete for the same land and the environment is deteriorating. Here, the main problem at present is not to conserve rare or endangered species – they are gone anyway – but rather to reverse the process of deterioration by establishing a restoration trend. This may well use the requirements of focal species to develop large-scale models and management guidelines to be applied to selected conservation areas (Lambeck 1997). In the lowland of Northern Italy, the Po valley, the natural landscape was represented by broad-leaved forest. Therefore, it may be assumed that nature conservation in this region largely coincides with conservation and reconnection of forest remnants.

The objective of this chapter is to present the applicability of focal species for developing an ecological network as a strategy for nature conservation and multi-functional land use in fragmented landscapes. Because of its urbanised and multi-functional land use, the Milan northern lowland has been used as a pilot study area (Bani *et al.* 2002). In addition to the forest ecological network implementation, we also present (a) an approach for wider-scale identification of forest core areas and (b) a tentative approach for an agricultural ecological network.

11.2 Geographical ecology

About ten years ago, a profound scientific and philosophical revision of the concepts on which conservation biology was traditionally based was started in the most developed areas of the world. It was perceived better and better that the traditional paradigm of nature conservation by means of a number of sites more or less representing the diversity of natural habitats had become largely unsatisfactory. In fact, while the total area officially protected under the establishment of any kind of parks and reserves had continued to increase over time an equally continuous loss of biodiversity was occurring, not only outside those areas but even well within their borders. Ironically, it was felt that conservation problems were becoming worse and worse while the total land coverage by protected areas was being increased. Clearly, it appeared that it was not possible to create an effective and lasting conservation system by means of an archipelago of small parks and reserves, not even though they were numerous, covering a significant area and including an adequate representation of all natural landscapes. This led to widening the scope, development of new concepts and the application of new techniques in conservation biology research.

Since MacArthur and Wilson (1967) published their classic work on biogeographical ecology and even more after the recent syntheses of Forman and Godron (1986) and Forman (1995) on landscape ecology and Hanski and Gilpin (1991) on metapopulation dynamics, biogeographers have been warning more and more about insular approaches to nature conservation. They have been claiming that an archipelago of small protected areas in an ocean of more and more intensive agriculture and more and more crowded towns will separate all residual wild animal and plant populations into a handful of more or less closed pockets. Such isolated populations may face very poor survival perspectives: they are prone to become genetically impoverished, owing to inbreeding, and to become more and more vulnerable, owing to external problems such as pathogens and other casualties (Soulé 1987). Genetic erosion is rapidly followed and paralleled by demographic erosion and may continue up to its extreme consequence, i.e. a local extinction. Furthermore, the final outcome of a process including several local extinctions is unavoidably a global extinction, a process that is occurring already for several species in different areas of our planet.

The local extinction process was carefully described and analysed by MacArthur and Wilson (1967) while studying oceanic islands. They noticed that species loss is faster and involves a higher number of species when a island is smaller and further from a continent. In fact, two inverse relations exist, between (a) the extinction rate and the area of an island and (b) the recolonisation rate and the distance of any island from the continent. Small islands lying in the middle of the ocean show high extinction rates and low recolonisation rates and hold a low species number for any taxon.

In the case of virtual islands, such as natural landscapes isolated within a large man-dominated area, the isolation is less extreme but, on the other hand, there is also a clear-cut margin effect of the surrounding anthropic matrix (Meffe and Carroll 1994). The major developments in this matrix are the growing isolation through larger and larger inaccessible areas and fragmentation through an increasing number of human-made barriers. This leads to an increasing number of isolated areas of which many are too small for the existing populations.

11.3   The ecological network

To reverse this trend, one should forget the classic static paradigm of protected areas seen as an archipelago and to replace it with a new paradigm seeing these institutions as *core areas* of an ecological network covering a whole region, a state or even a continent. To this end, both in Europe (Bennett 1991; Bischoff and Jongman 1993), and in North America (Noss 1992; Soulé and Terborgh 1999a), it is felt more and more that ecological corridors and buffer zones should be carefully planned to connect core areas and, respectively, to provide a useful transition to anthropic areas. As a whole the ecological network (core areas, corridors and buffer zones) should increase the inclusive protected area to cover much more than the present 10% (a figure referring to Italy), reaching about 25–30% of the whole country (Bennett 1991; Noss *et al.* 1997). The main difference between the traditional protected areas and those newly selected as corridors and buffer zones is that the latter should be subjected to multiple use, i.e. should also be used for a number of compatible economic activities such as pastoralism, hunting, fishing, ecotourism, sustainable forestry and other kinds of extensive agriculture. This is a quite new conservation concept, which is becoming more and more accepted. These new concepts were largely presented and explained recently under the title *Continental conservation* (Soulé and Terborgh 1999a). The editors state in their preface that the immediate objective of their book is

> to produce – and then implement – map-based proposals for an effective network of nature reserves through North America. The elements of the networks include large, wilderness-like core areas linked by landscape corridors to facilitate natural flows. The network should be buffered, where appropriate, by lands that may also serve economic objectives.

11.4   New concepts and new technology

A recently developed mean of selecting areas for conservation is by using focal species (Lambeck 1997; Noss *et al.* 1997). This is similar to the classical approach of umbrella species (Wilcox and Murphy 1985), except for the fact that (a) it uses as a tool a small group of species rather than a single one, and (b) it places a greater emphasis on relationships between landscape pattern–processes and community ecology parameters.

In addition, it should be stressed that new conservation concepts at a continental or local scale can only be developed with the help of a Geographical Information System (GIS). GIS is a computer-based system to process geographical data, often put into operation in conjunction with other technological systems such as remote sensing (see, for example, chapter 5). The two techniques in conjunction may allow land cover to be analysed at a high resolution (e.g. 30 m × 30 m) and in this way may help to understand spatial ecological relationships much more clearly than was ever possible in the past.

GIS and remote sensing techniques made our approach possible and, in conjunction with the classical wildlife census methods and focal species concept, helped to develop a strong methodological instrument for building ecological networks.

11.5   The Italian Northern Lowland pilot project

Our study area for this forest network project lies north of Milan (Figure 11.1) over 2500 km². It is bordered by the lower limit of the pre-Alps to the north, by the historical Milanese channels to the south, and by the Ticino and Adda rivers on its western and eastern borders, respectively. Human population density is very high, reaching 2000 people per square kilometre. However, about 30% of this area is still wooded and subjected to legal protection, mostly under the status of Regional Park.

To ascertain the possibility of and criteria for conservation planning in this area, we first of all (a) assessed the land use pattern at a sufficiently fine grain, then (b) determined which bird species found within wooded areas may be suitable as focal species. Once we had accomplished these two basic data-dependent steps, we (c) related land use pattern to focal species distribution and abundance to produce a suitability map; and finally (d) designed a network of corridors suitable for focal species. We performed these steps as follows (see Bani et al. 2003).

(a) To assess the pattern of land use, our first step was the interpretation of a satellite image of our study area. This was taken on 17 May 1997 by the TM sensor of the Landsat 5 satellite using 7 spectral bands at a resolution of 30 m × 30 m. Interpretation and georeferencing of this picture was performed by using ENVI 2.7 software (Research Systems Inc.). The resulting types of land use are: built-up areas (32.6%), agricultural areas (25.7%),

FIGURE 11.1.
Location of the study areas.

broad-leaved woodland (19.0%), mixed and coniferous woods (4.5%), locust-tree (*Robinia pseudoacacia*) woods (2.9%), shrubs and prairie (12.3%), water (2.3%), barren lands (0.6%).

(b) Because of the restricted number of species present and the multi-functionality of the landscape, the best candidate focal species are the extant forest vertebrates. Among them, forest birds can be successfully used. We then developed a procedure to select forest focal species. We did this starting from our bird census database by choosing forest birds positively correlated with broad-leaved forest and also showing a significant mutual correlation. The result of this procedure was the selection of the following four species: green woodpecker (*Picus viridis*), great spotted woodpecker (*Picoides major*), marsh tit (*Parus palustris*) and nuthatch (*Sitta europaea*).

(c) To relate land use pattern to focal species distribution and abundance, we related habitat variables to focal species abundances by applying a correlation analysis followed by a multiple regression. This resulted in a suitability map that might be used to identify the core areas to be connected (Figure 11.2).

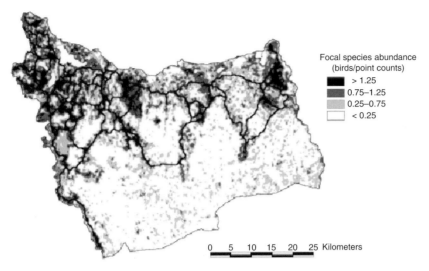

Focal species abundance
(birds/point counts)

■ > 1.25
■ 0.75–1.25
▨ 0.25–0.75
□ < 0.25

0   5   10   15   20   25 Kilometers

FIGURE 11.2.
Northern Lowland study area: potential corridors (bold lines into the map)
connecting core areas (highest focal species abundance areas) for focal bird species.

(d) To identify the possible connections, we drew the shortest possible lines
   connecting separate core areas through a path showing the best available
   landscape. This resulted in the identification of potential corridors for
   forest focal species within the study area (Figure 11.2).

The whole process is summarised in Figure 11.3.

11.6   Forest core areas identification on a larger scale

Our next purpose was to identify forest core areas in a wider region to
be able to build up more forest networks within the whole of the Lombardy region (Figure 11.1). To this end, we used our bird census database for the whole
region (about 24 000 km^2) and we also identified different forest types characterised by different forest focal species.

The census database of the whole Lombardy region (1992–99) consisted of
2795 point counts. The total number of birds counted was about 114 000, pertaining to 178 species. As the main purpose of our work was to identify forest
core areas in quite diverse landscapes, we divided the region into faunal districts, representing homogeneous faunal areas. To do this we chose simplified
hydrographic basins as primary units, and grouped them on the basis of their
faunal composition. Subsequently, we accomplished a basin simplification by
starting from the regional CORINE layer and (a) intersecting it by the 100–200–
300 m isohypses to obtain smaller areas in the lowland, (b) grouping the smallest basins into larger homogeneous and contiguous areas in the mountains.

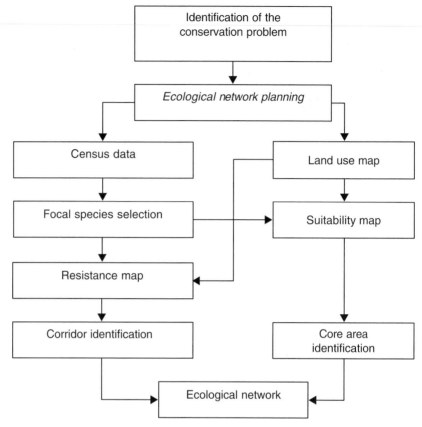

FIGURE 11.3.
Schematic representation of the procedure used for the Northern Lowland pilot project.

Finally, we performed a simplified basin cluster analysis based on their faunal composition. All of this resulted in the identification of the six faunal districts shown in Figure 11.4. At this point, for each district and for the whole of Lombardy as well, a small set of focal species has been selected for the two main woodland types, coniferous and broad-leaved woodland according to the regional CORINE cover map. We did this both at regional and at district level.

At the regional level, the appropriate set of focal species was chosen on the basis of two rarity and one habitat selection attributes. These are: (a) frequency (presence in 0.5–10% of the point counts), (b) abundance (species covering 0.01–1% of the total bird number), (c) woodland habitat selection (weighted mean of forest cover beyond 30%, with a standard deviation not higher than 40%). In addition, we excluded those species clearly pertaining to local non-woodland habitats, such as rocks or water, although appearing in the woodland domain. At this point, we needed an analysis to discriminate species subsets

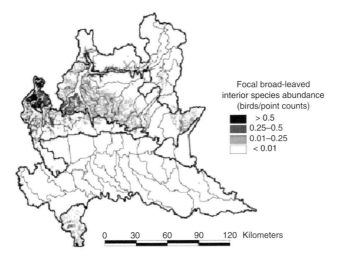

FIGURE 11.4.
Simplified basins, faunal districts and abundance of focal broad-leaved interior bird species in Lombardy (*Parus palustris*, *Sitta europaea*). Bold lines represent district limits, thin lines represent basin limits.

on the base of their ecological correlates. We did this in two different ways (see Jongman *et al.* 1995): (1) by a multi-dimensional scaling analysis, which resulted in the grouping appearing in Figure 11.5A, and (2) by a cluster analysis (Figure 11.5B), which offered quite similar results. We identified four main subsets related to (a) broad-leaved woodland, (b) interior broad-leaved woodland, (c) broad-leaved woodland margins and (d) coniferous woodland.

An example of the possible outputs of this kind of approach is shown in Figure 11.6 in which the Kernel procedure (Worton 1989), usually adopted to evaluate animal home ranges, has been used at a different scale to draw a weighted quantitative map of focal species abundance. This is clearly correlated with forest core areas and may represent an optimal base from which to plan ecological forest networks. Both Figure 11.4 and Figure 11.6 show that the focal species abundance is lower where the forest is more fragmented.

11.7 A case study in an agricultural landscape

Ecological networks may also be established in areas even more heavily exploited and almost devoid of forest. However, these may not be forest networks but rather other types of networks resulting from landscape marginal elements affecting matrix quality. As an example, we examine the open area just south of Milan, presently protected in the form of a large agricultural park (Parco Agricolo Sud Milano) (Figure 11.1). Here, the total forest coverage, also including minor, low-quality woodland and even poplar

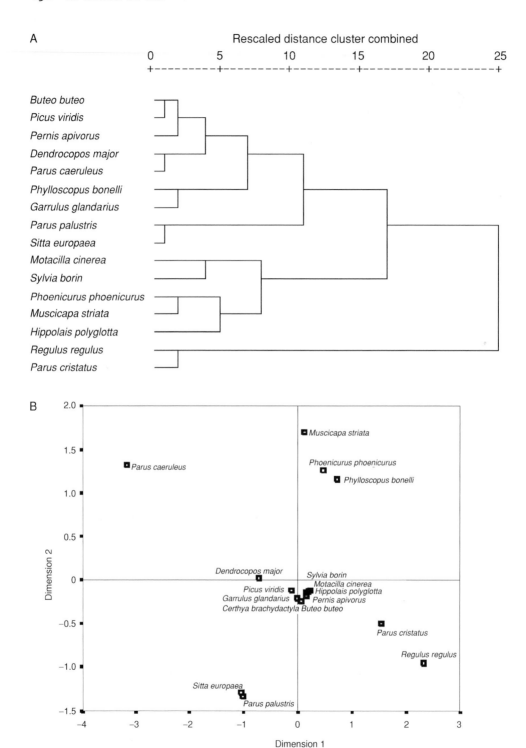

FIGURE 11.5.
Focal species assemblages identification by means of multi-dimensional scaling
(A) and cluster analysis (B). Matrix distance, city block; dendrogram shows complete
linkage.

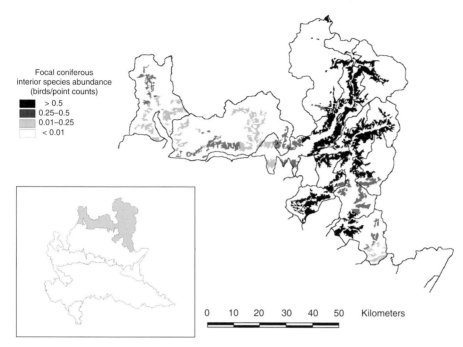

Focal coniferous
interior species abundance
(birds/point counts)
> 0.5
0.25–0.5
0.01–0.25
< 0.01

0   10   20   30   40   50   Kilometers

FIGURE 11.6.
Simplified basins and abundance of interior coniferous focal species (*Dryocopus martius, Turdus philomelos, Regulus ignicapillus, Parus montanus, Parus cristatus, Certhia familiaris, Loxia curvirostra, Pyrrhula pyrrhula*) in the alpine district.

plantations, does not exceed 9% of the whole area, and almost half of it is in fragments smaller than 20 ha. To design an ecological network in this area the main problem is to ascertain which particular type of landscape should be accepted as a desirable compromise between biodiversity conservation needs and a reasonable type of economic land use in a 430 km² area where about 800 farms are located. Clearly, the best candidate to this end appears to be a more or less traditional, extensive agricultural landscape.

Therefore, we selected from our database another species assemblage covering the agricultural landscape. To this end, we first performed a multi-dimensional scaling analysis and a cluster analysis. The procedure produced the four groups appearing in the rows of Table 11.1. At this point, we looked for significant correlations of each group to (a) the presence of woodlots within the agricultural landscape, (b) distance from a hedgerow 30 m wide, (c) distance from a hedgerow 12 m wide, (d) distance from a hedgerow 3 m wide, (e) presence of open farmland only (columns of Table 11.1).

It appears that some of the so-called agricultural species are significantly correlated mostly with trees (e.g. wren (*Troglodytes troglodytes*) and blue tit (*Parus caeruleus*) groups), others with shrubs (e.g. blue tit and especially turtle dove

Table 11.1. *Correlation matrix between agricultural bird groups and landscape parameters*

P is the significance level that results from a Spearman correlation analysis.

Bird group		(a) Woodlots	(b) Distance from hedgerows 30 m wide	(c) Distance from hedgerows 12 m wide	(d) Distance from hedgerows 3 m wide	(e) Farmland
(1) *Troglodytes troglodytes*		0.356	−0.154	0.091	−0.003	0.044
	P	0.000	0.009	0.124	0.964	0.459
(2) *Parus caeruleus*, *Oriolus oriolus*, *Picoides major*		0.334	−0.197	−0.121	−0.063	−0.102
	P	0.000	0.001	0.04	0.287	0.085
(3) *Streptopelia turtur*, *Lanius collurio*, *Hippolais polyglotta*, *Muscicapa striata*, *Columba palumbus*		0.144	−0.249	−0.282	−0.183	0.03
	P	0.014	0.000	0.000	0.002	0.614
(4) *Motacilla flava*, *Saxicola torquata*		−0.197	0.009	−0.055	−0.131	0.254
	P	0.001	0.886	0.354	0.027	0.000

(*Streptopelia turtur*) group), and some others with open farmland without tree or shrub cover (e.g. yellow wagtail (*Motacilla flava*) group).

This suggests that the agricultural landscape is not a homogeneous habitat block but rather a discrete mosaic of different types of patches such as different crop fields, channels bordered by trees, double channels bordered by hedgerows, etc. Therefore, to achieve better connectivity within an agricultural landscape, it will be necessary to evaluate the grain and composition of the elements to create different matrix or/and linear connection types for different species.

11.8 Conclusions

The case studies outlined here suggest that, from a theoretical point of view, the establishment of an ecological network represents something quite feasible. However, there are a number of problems in doing this in the real world.

The first problem is that of deciding which kind of habitat or landscape should be conserved in a particular geographical region. This may seem obvious, but in fact it is often much more based on politics and human sciences than on bio-geographical and bio-historical considerations. A second problem is to decide which particular kind of organism we should use as an ecological indicator to plan the ecological network. The decision that has to be taken is about the level of conservation (or restoration) to be achieved and therefore which particular focal species should be used to do this in the best way. It is clear that the technical content of such a decision is rather small in comparison with its political implications.

As any network is supposed to cover a very significant portion of the area in which it is placed, it will be impossible to implement such projects unless national or pan-continental economy strongly includes these goals within political trends.

For this reason, an effective ecological network implementation is not a project that may be easily put into practice, but a complex system. In conclusion, the concept of ecological networks offers possible mitigation of dramatic problems by breaking the anthropic ecosystem and anthropocentric concerns. First of all it might help to solve the problem of biodiversity loss, but it also introduces new problems, primarily the need for a careful revision of the present economic concepts.

12

Connecting corridors: implementing metropolitan greenway networks in North America

12.1 Introduction

Greenways receive wide attention as landscape connectors for conservation, recreation, transportation, and neighbourhood enhancement. Hundreds of greenway projects are either complete or in progress in cities across North America; nearly all of the largest cities are planning for interconnected greenway networks (Grove 1990). Greenways are linear open spaces along natural or human-made features such as rivers, ridgelines, railroads, canals or roads. They are planned, designed and managed to connect and protect ecological, scenic, recreational and cultural resources. A greenway may include trails or it may be a conservation corridor without recreational access. Greenways have a number of other names (Ahern 1995; Cook and Van Lier 1994), each implying a slightly different focus, such as extensive open space systems, ecological networks, ecological infrastructure, wildlife corridors, or habitat networks.

Consider this scenario: a greenway network plan has been created for a metropolitan region. A number of entities have been involved in the effort, including the general public, governmental agencies, and non-governmental organisations. Given this, the greenways project still needs a structure in which to work – an organising method or institutional arrangement for implementation. Therefore, one main question frames this study: how have greenway projects been implemented in North American cities? This research focuses on interconnected greenway networks, rather than on individual corridors. (The term 'network' is used here to mean a meshed fabric of nodes and connecting corridors.)

'We frequently perceive crises of governance where nothing seems to get done. Even after public policy is developed, 'implementation' by definition does not actually get things done, but merely lays the groundwork...' (Quayle 1995). This chapter considers this challenge. It examines the greenway implementation structures for: Minneapolis, Minnesota; Portland, Oregon; Toronto, Ontario; Chattanooga, Tennessee; and Chicago, Illinois, and suggests implementation models that can be applied to greenway networks in other cities. From the case studies examined in this chapter, it appears that two main factors are particularly critical: motivating greenway networks out of a diverse range of objectives; and creating an effective institutional structure.

12.2   Background

Greenways are unique landscapes. As long linear features, they are often developed on a complex mosaic of public and private lands. They traverse property lines, natural features, and other public infrastructure. They transcend jurisdictional boundaries and sometimes even state lines. Greenway development usually includes aspects of both creating new spatial forms and protecting existing resources. For these reasons, implementation is difficult and specialised. Cross-jurisdictional co-operation is imperative. 'In terms of greenway conservation, the consequences of isolated, local institutions is a cacophony of competing, conflicting land use policies, none of which takes into consideration the greenway in its entirety. If no attempts are made to co-ordinate these conflicting practices, degradation or even loss of the greenway and its dependent resources is possible' (Hoover and Shannon 1995). Given this unique scope, complexity and audience, it is important to understand how these networks are realised.

According to ecologists, it is imperative to implement natural areas in a connected web at large scales. Conservation efforts are still relatively local, emphasising isolated preserves, where protected lands are a fragmented patchwork, threatened by exotic species, edge effects, and increasing human disturbance and encroachment (Dale *et al*. 1999). 'The elements of the solution are known: bigness and connectivity...On-the-ground realization of a program of large core areas and landscape connectivity will, however, require research, planning, and bold advocacy at unprecedented scales' (Soulé and Terborgh 1999).

While the outcomes of large-scale connected open space are ecologically desirable, the political and social scale at which this connectivity is accomplished is less certain. The debate over an appropriate scale for public projects is not a new one and the efficiency of 'regionalism' is controversial. Detractors have claimed, for instance, that regional government is not a remedy for spatial problems (Siegel 1999) and that, although some problems are best served by large metropolitan governments, most problems are best scaled at more local levels. 'Given certain levels of information, technology, communication, and certain patterns of identification, a scheme might be imagined which had an appropriate scale of public organisation for each different public good'

(Ostrum *et al.* 1961). It can quite plausibly be argued that open space protection (as one public good) is an area where planning and implementation at the regional level is effective, in both ecological and humanistic terms. Apart from the ecological benefits, people may actually perceive linear corridors in larger frameworks than they do other land uses. Arguing for the perceptual benefits of a regional approach Lynch states: 'I mean to take the peculiar position that the experiential quality of the environment must be planned for at a regional scale...since people now live their lives at that scale' (Lynch 1976).

Previous research provides valuable knowledge about greenway design and planning, particularly focused on historic precedents, the planning process, and citizens' mobilisation (Little 1990; Flink and Searns 1993; Ryan 1993; Ahern 1991). Specific greenway planning issues have received close attention: economic impacts (National Park Service 1991, 1992); ecological benefits (Smith and Hellmund 1993; Ahern 1991; Labaree 1992); conceptual approaches and typologies (Ahern 1995; Turner 1995); conflict resolution (Moore *et al.* 1992); the public's willingness to pay (Lindsey and Knaap 1999), use characteristics (Lindsey 1999; Gobster 1995; Foruseth and Altman 1991); and users' perceptions (Moore and Graefe 1994).

The linkage of greenway corridors into larger networks has received some attention in the planning literature. Milos *et al.* (1995) developed methods and criteria for incorporating individual corridors into regional greenway networks. Their work suggests a landscape planning methodology for integrating local 'stand-alone' greenway corridors into a regional greenway infrastructure. However, little research has focused closely on the greenway implementation process, partly because many of the regional greenway systems have only been planned over the past decade and systematic implementation has only begun in many cases (Erickson and Louisse 1997). Ahern (1994), in an article on greenways as ecological networks in rural regions, defined a set of implementation tools and techniques, including acquisition, regulation and voluntary options. Hoover and Shannon (1995) also identified the opportunities for collaboration and the implications for those opportunities, using the Tug Hill region of New York State as a case study area. Likewise, Quayle (1995) reports on the decision-making process for implementation of a greenway system in Vancouver, British Columbia. Taylor *et al.* (1995) analysed the evolution of Canadian greenway networks, focusing on the approaches used in four metropolitan case studies.

Much has been written on the implementation of other public facilities. A body of literature in public administration and political science focuses on the science of how policies and programmes are implemented (Lindblom 1959; Cates 1979; DeYoung and Kaplan 1988; Sanger and Levin 1992; Ostrum *et al.* 1961; Behn 1988; Goggin *et al.* 1990). Similarly, intergovernmental relations in open space planning and implementation have been examined (Lowry 1998; Press 1998, 1999). The theoretical frameworks developed in the implementation literature are useful in analysing how metropolitan greenways develop from vision to reality. In this chapter, the implementation literature is applied to greenway planning, particularly as it relates to objectives, scale, government roles, and institutional structures.

12.3   Case studies

Figure 12.1 shows the five metropolitan areas that form the basis of this study. These cities were chosen because they: (1) plan a network or system of linear open spaces; (2) include multiple jurisdictions and large landscape units; (3) focus on multiple functions and objectives; (4) represent projects currently being implemented; and (5) serve as oft-cited examples of greenway success stories (Little 1990; Flink and Searns 1993; Smith and Hellmund 1993). Document analysis and a key informant interview process were used to gather comparative data for each case study. In selecting interviewees, referrals were sought from people knowledgeable about greenways work, both at a broad level and in the specific case study sites. People from public agencies and from private groups (such as not-for-profit collaborators) were interviewed for each case study location.

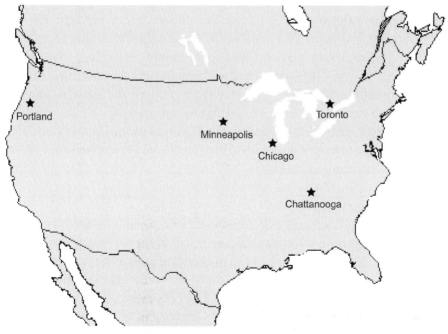

FIGURE 12.1.
Location of the case study areas.

As shown in Table 12.1, greenway plans for the five case studies were completed in the 1990s. They vary from 120 to 1477 proposed kilometres of new greenway corridors. All five are organised around natural water bodies. They focus on river corridors and, in the case of Chicago and Toronto, the Great Lakes. Four of the five use county lines as project boundaries. In contrast, it is interesting that Toronto's greenways vision is structured around the geologic features of the Greater Toronto bioregion – Lake Ontario, the Niagara Escarpment and Oak Ridges Moraine. Important similarities and differences affect planning and implementation of these greenway networks. First, of course, size and scope vary dramatically, from the smaller city of Chattanooga to the large metropolitan regions of Chicago and Minneapolis. Table 12.1 illustrates the numbers of jurisdictions involved in each project, ranging from three to nine.

12.3.1 Chattanooga

In Chattanooga, Tennessee, several grassroots efforts came together at nearly the same time to focus on greenways, leading to a Chattanooga – Hamilton County greenways network. The City established a greenways advisory board in 1989 to develop strategies for a city-wide system and to implement the North Chickamauga Creek Greenway Plan. The Board, the City, and the Planning Commission asked the National Park Service's Rivers, Trails and Conservation Assistance (RTCA) Program to help develop strategies for implementation. *Greenways for the Southeast Tennessee River Valley* was completed in 1995 (National Park Service 1995), identifying 30 corridors in an eight county area of Tennessee and Georgia. Through workshops in each county during the early 1990s, nine corridors were prioritised and active citizen-led efforts began on five corridors. These groups, composed of citizens and non-profit organisations, act as advocates for individual greenway corridors. The Trust for Public Land provides oversight as a metropolitan coordinator. It organised a Chattanooga Greenways Task Force that represents all five groups, co-ordinates their efforts, and advises on funding allocation. Elected officials control the purse strings.

The spine of the Chattanooga greenways network is the Tennessee RiverPark, a greenway along both sides of a 16 km stretch of the Tennessee River, planned throughout the 1980s (Figure 12.2). In the RiverPark planning process, including hundreds of public meetings, one of the main themes that emerged was the public's desire for river access. River Valley Partners, a private non-profit group, was formed to acquire land and to implement a Tennessee RiverPark Master Plan. It obtained private funding to begin the RiverPark, now a stunning stimulus to redevelopment of this former rustbelt city (Ivy 1992; Jacobson

Table 12.1. *Comparison of greenway factors for five case study sites*

Metropolitan area	Name of plan/date	Jurisdictions	Length	Waterbody/natural features	Boundaries
Chattanooga	Greenways for the Southeast Tennessee River Valley (1995)	8 counties	Over 120 km proposed	Tennessee River and tributaries	County lines
Chicago	Northeast Illinois Regional Greenways Plan (1992)	6 counties	1477 km proposed (another 1087 km existing)	Chicago River and tributaries; Lake Michigan	County lines
Minneapolis	Metro Greenprint[a] (1997)	7 counties	Unknown	Mississippi River and tributaries	County/City lines
Portland	Metropolitan Greenspaces Master Plan[a] (1992)	3 counties (coordinated with 1 county in WA state)	563 km proposed	Willamette and Columbia Rivers and other streams and natural areas	County lines
Toronto	Lake Ontario Greenway Strategy (1995)	9 regional jurisdictions	901 km proposed; 201 km Waterfront Trail	Lake Ontario, Niagara Escarpment and Oak Ridge Moraine	Greater Toronto Bioregion

[a] These plans target public open space more generally, but include greenway corridors as an important component.

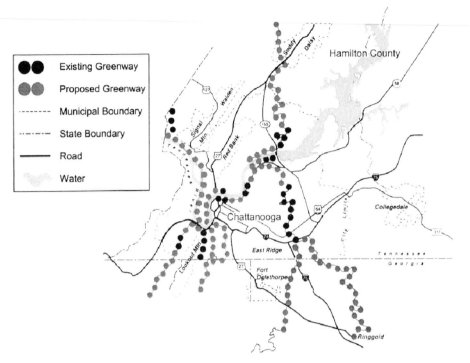

FIGURE 12.2.
The Chattanooga – Hamilton County Greenway System (adapted from Trust for
Public Land 1995).

1997). Chattanooga's mayor fully supported the effort by empowering others
to facilitate the project. Citizen support was overwhelming. The Lyndhurst
Foundation, a Chattanooga philanthropic organisation, was also instrumental
in supporting the vision.

12.3.2 Minneapolis/St Paul

Minneapolis and St Paul, Minnesota (the Twin Cities), are well known in
the USA for their extensive parks, trails and open space system. Although the
metropolitan region has a very new planning infrastructure for regional greenways, there has been strong grassroots support for parks and open spaces since
the mid-nineteenth century. This legacy can be traced to the plans of influential
designers Horace Cleveland and Theodore Wirth. However, there has also been
a reluctance to form partnerships within larger regional contexts; a strong history of opposition to regional authority, planning and implementation exists
(Orfield 1997).

The Metropolitan Council of Minneapolis and St Paul was created in 1967
to coordinate development in a seven county area with 140 jurisdictions
(Metropolitan Council 1995). Over that time, the Metropolitan Council and

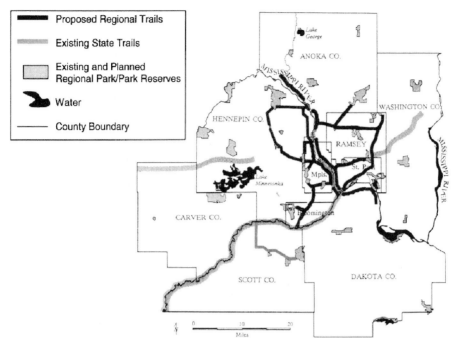

FIGURE 12.3.
Proposed regional trails in the region of Minneapolis and St Paul.

local counties developed a regional open space system including 18 210 ha of parkland and 128 km of public trails (Greenways and Natural Areas Collaborative 1997). Figure 12.3 depicts the trails proposed in recent Metropolitan Council (1991) plans for the seven county region. However, up to the late 1990s, the Metropolitan Council had not achieved the level of coordination that is necessary for regional greenways visioning. Meanwhile, urban land has expanded by 25% for every 10% increase in population in the Twin Cities (Cisneros 1995). In addition, population is expected to grow by 650 000 in the next two decades (Greenways and Natural Areas Collaborative 1997).

More recently, a Metropolitan Greenways Program within the Minnesota Department of Natural Resources is creating collaborative greenway planning opportunities across the Minneapolis/St Paul region. The Program, begun in 1997, addresses natural areas, open spaces and greenways. The Program, while new, has a focused direction toward a 'bottom-up' approach, with local units of government initiating greenway connection projects. It provides technical assistance, planning grants and land protection funds.

Even though this regional greenways vision is newly formed, greenways have been steadily implemented in the Twin Cities region. Over US$500 000 in matching grants have been awarded to implementing agencies. Greenways are being created within city and county boundaries, and in some cases connect across jurisdictional boundaries. In this metropolitan area, a strong sense of public priority, over many decades, is more notable than the leadership of one individual with an expansive regional vision and a set of programmes. For instance, in Hennepin County, west of Minneapolis, Hennepin Community Works is a progressive programme that incorporates infrastructure, transportation, and overall system needs into one vision. This innovative programme was created by a Parks and Public Works Commission for Hennepin County. It co-ordinates how communities within the County invest in infrastructure, public works, parks and natural systems by improving existing implementation systems. Part of this vision is to strengthen communities through the types of connection that greenways provide (Bryant 1998).

12.3.3 Chicago

From 1970 to 1990, the Chicago, Illinois, metropolitan area grew only 4% in population, but land used for housing and commercial development increased by 45% and 74%, respectively (Cisneros 1995). Population in the outer counties of the Chicago region is projected to increase by 70–100% by 2020. Responding to this sprawling development pattern, the Northeastern Illinois Regional Greenways Plan was developed by the Northeastern Illinois Planning Commission (NIPC), a regional agency, and Openlands Project, a regional non-profit group (Northeast Illinois Planning Commission and Openlands Project 1992). Completed during a period of political decentralisation in the Chicago region, this is the first large-scale metropolitan greenway plan in the USA. One of the plan's major goals is filling the gaps between existing greenway corridors (Figure 12.4). This is a highly complex project, involving 260 local jurisdictions and more than 150 Park Districts and Forest Preserves. Despite planning activity at a regional level, including planning charrettes in each of the six counties, there is no central agency that implements the greenway plan. Rather, local jurisdictions implement projects, with the State of Illinois primarily administering funds. However, Northeast Illinois Planning Commission and Openlands Project serve as important facilitators.

Openlands Project strategically pursues and facilitates several greenway projects each year in order to implement the regional plan and to insure that greenway efforts stay active. For instance, the Salt Creek Greenway, built in phases, is a connector trail that passes through 35 jurisdictions. The co-ordination effort for this project was substantial, as it involved property belonging to Forest Preserves, Park Districts and Cook County. The regional greenways plan was also incorporated into the State of Illinois transportation enhancement guidelines, which provided early implementation success.

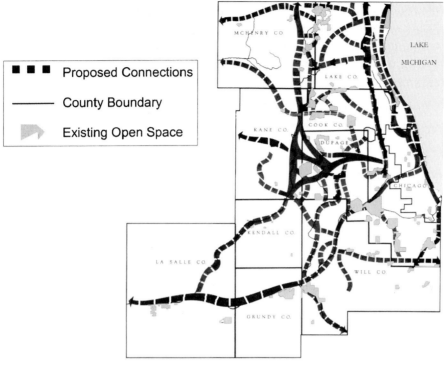

Proposed Connections

County Boundary

Existing Open Space

FIGURE 12.4.
Concept plan for Chicago Open Space Connections (NIPC 1992).

Even with these successes, the lack of a regional implementation structure presents some challenges. Although there is a greenway priority list within the Greenway Plan, many of those projects listed as priorities have local issues that stall implementation. In the meantime, other greenways are pursued, not necessarily in an optimum sequence. Perhaps this incremental and non-linear process is inevitable in a city as large and diverse as Chicago. It also highlights that, although the plan was adopted at a regional level, the implementation is not accomplished regionally, but rather locally. A 1994 State of the Greenways Report (Northeast Illinois Planning Commission and Openlands Project 1994) showed, however, that all six counties and the City of Chicago have actively implemented parts of the 1992 plan. Chicago's big vision has stayed alive and active over the past decade. A 1997 plan update triples the size of the proposed network and doubles the trail component. It includes a separate water-based trail system and delineates a focused action plan for implementation (Northeast Illinois Planning Commission and Openlands Project 1997).

Other important participants in Chicago are the National Park Service and Friends of the Chicago River, which worked together to create a Chicago River enhancement plan. In 1992 Friends of the Chicago River conducted a study of the use and perceptions of the river (Wallin 1995). This yielded important information about inequitable access to the river, among the diverse populations residing near the river. The Chicago River was not being pursued as a greenway resource prior to this project.

12.3.4 Portland

Portland, Oregon, has been facing the competing pressures of growth and land preservation for decades. Two Oregon governors were early promoters of the greenway concept, and Portland's successes are outgrowths of that foresight. The Willamette River corridor was declared a greenway in 1967, long before many of the current initiatives began. More recently, Metro, Portland's regional government, has sought to protect and set aside open space, natural areas, habitat, and recreation areas in the entire region (Metro 1992). A bond measure passed in 1995 helps fund these efforts. Portland's Metropolitan Greenspaces Master Plan (Figure 12.5) is a national model of co-operative planning involving more than fifty government agencies, many conservation organisations, businesses, friends' groups, and interested citizens in Oregon and Washington. The structure of these planning groups was organised into three branches addressing the technical, political, and community issues of the

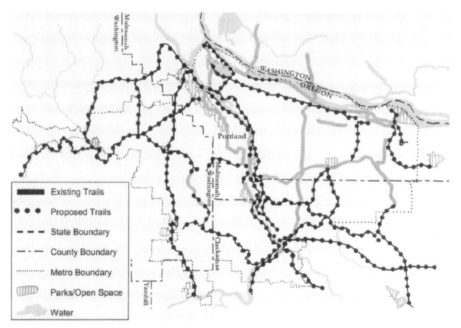

FIGURE 12.5.
Concept plan for Portland Open Space, Parks and Trails (adapted from Metro 1999).

greenway initiative. The final master plan contained results derived from the co-ordination of all three groups.

Subsequently, 22 of 24 cities within the boundaries of Metro and all four counties have passed resolutions of support for the Metropolitan Greenspaces Master Plan since 1990 (Porter 1995). Metro is the planning coordinator and key implementer of the programme. It has the authority to buy land and to implement certain planning and growth management measures on a region-wide basis. Simultaneously, Metro co-operates extensively with local jurisdictions. Metro's goals are to (1) co-ordinate the efforts of government and citizen groups to create a regional open space system, (2) negotiate public access agreements, and (3) own and operate some of the acquired lands. For the regional trails and greenways system, composed of six target corridors, Metro co-ordinates planning, funding, acquisition, design, development, construction, operations, and maintenance. About 2400 ha of land have been acquired since passage of the 1995 bond measure. This includes land in twelve regional natural areas and the six greenway corridors (Metro 1999).

12.3.5 Toronto

The Waterfront Regeneration Trust (WRT) is the main co-ordinating mechanism for greenways in the greater Toronto, Ontario, metropolitan region. WRT was created by an act of the legislature in 1992 and in 1999 became a private non-profit organisation. Its predecessor, The Royal Commission on the Future of the Toronto Waterfront, conducted a study in the late 1980s, which proposed developing greenways within a 10 360 km^2 area known as the Greater Toronto Bioregion (Royal Commission on the Future of the Toronto Waterfront, 1992). Although the WRT has no power or authority to set policy, it does have the ability to hold land. Its mission is to look at the interface of land and water and to support watershed groups that are conducting land-use initiatives. It acts as a facilitator to bring parties together, develop agreements on waterfront objectives, co-ordinate funding proposals, report on progress and challenges, and promote appropriate conservation and use of the waterfront (Waterfront Regeneration Trust 1995). The area of focus is primarily the Lake Ontario waterfront along its entire north shore; however, WRT collaborates on inland greenway projects as well.

As in the Portland case, strong leadership was critical in Toronto. The former Mayor of Toronto, later a federal politician, was the visionary leader behind the Toronto waterfront focus and was extremely influential in the success of this project. He provided the leadership and enthusiasm that sparked the entire project, especially its emphasis on ecosystems and their accessibility by people.

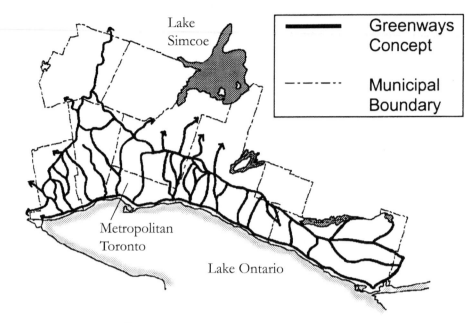

FIGURE 12.6.
Greater Toronto Greenways and Trails Concept (adapted from the Royal
Commission on the Future of the Toronto Waterfront 1992).

An emphasis on environmental quality led to the greenways theme. Comparing examples of metropolitan greenway networks across Canada, Taylor *et al.* (1995) use Toronto as an example of a progressive approach:

> Collectively these examples [*Calgary, Saskatoon, Toronto, Ottawa*] demonstrate an evolution from an urban design approach that attempted to impose both landscape form and land use function, to an ecologically based planning approach that addresses natural factors, connections between natural and urban systems, public participation and support, and innovative government involvement.

Toronto's greenway system (Figure 12.6) is planned at a regional level but decentralised to local jurisdictions for implementation. The Toronto and Region Conservation Authority, which plans and implements conservation activities in the nine watersheds (14 000 ha) of the Toronto region, is a key participant in developing inland greenways along stream corridors. The greenways network along Lake Ontario is based on the Lake Ontario Greenways Strategy (LOGS), completed through the cooperation of a steering committee of sixty people (Waterfront Regeneration Trust 1995). The LOGS sets a broad conceptual framework for completion of the greenway, since the responsibility for implementing greenways does not rest with any single agency. Rather, each of the agencies, municipalities, and community groups reviews its own area of responsibility and actions to ensure that it is contributing in a positive way to achieving the vision and objectives of the Strategy.

Working with Greenways Advisory Committees in separate geographic areas, WRT helps organise schedules, priorities, and deadlines. Federal and provincial agencies, conservation authorities, and municipalities are encouraged to incorporate the LOGS into their planning, regulation, and other activities. The academic community is asked to assist in the research and monitoring. LOGS implementation is based on the following mechanisms:

- planning and regulatory mechanisms, including environmental impact assessments, waterfront legislation, watershed strategies and plans, remedial action plans and integrated shoreline management plans
- stewardship, including management of public land, land acquisition by public agencies, and landowner contact
- funding incentives, including matching grants and attracting private funds to waterfront projects

The plan outlines a two-tiered implementation strategy based on arterial greenways along major geographic features and local connectors within each community.

12.4 Cross-case themes

Greenway networks are clearly difficult to implement. Bardach (1977) writes on implementation:

> It is hard enough to design public policies and programs that look good on paper. It is harder still to formulate them in words and slogans that resonate pleasingly in the ears of political leaders and the constituencies to which they are responsive. And it is excruciatingly hard to implement them in a way that pleases anyone at all, including the supposed beneficiaries or clients.

A typical US metropolitan area encompasses 85 units of government, including, on average, two counties, 13 townships, 21 municipalities, 18 school districts, and 31 special districts (such as fire, sewer, or library districts) (Boyne 1992). Superimposing a spatial pattern of protected open space over this complex administrative structure is always daunting.

Quayle (1995) succinctly describes the array of road-blocks to getting things done, from her experience with the Vancouver, British Columbia, greenway project: a culture of conformity and compliance; the political and bureaucratic fear of change and the unknown; lack of a sustainable vision for the city; and visual, social and ecological illiteracy. All of these challenges and more are represented to some extent in these five case study sites. It appears that at least two main factors are critical: motivating greenway networks out of a diverse range of objectives; and creating effective institutional structures.

12.4.1 Multiple objectives

For contemporary greenway projects, implementation may depend to a large extent on the objectives that the projects are attempting to serve. Evidence from the cases shown here suggests that successful greenway projects are fuelled by multiple objectives and by objectives that diverge from the traditional parks and recreation foci. Table 12.2 depicts the main objectives for the cases examined in this study. The most common objectives are recreation, conservation, and neighbourhood enhancement. In most places, the protection of natural features is a particularly strong motivator. Other important objectives across this sample are environmental education, non-motorised transportation, and water quality protection. Some of these greenway projects have focused closely on water management objectives more recently, including the protection of stream corridors and watershed planning. Economic development and growth management issues are less likely to spur greenway planning at the outset. However, these are sometimes secondary effects that are later connected to implementation. Similarly, scenic quality and historic preservation are mentioned less often, but are sometimes important long-term benefits of greenways.

Keeping broad objectives visible and operational in the greenway planning process is vitally important, both for developing political will and for implementing meaningful projects. Portland demonstrated this by tying its open space programme to the issue of growth management and the protection of natural habitat. Portland's example is probably the most comprehensive of the cases studied here. Its greenway initiative is strongly tied to other plans for natural area protection and growth management. Portland's '2040 Framework' plan, developed by Metro in 1992, is the guiding document. It is designed to integrate land use, transportation, water and open spaces over a fifty year period (Porter 1995). The Metropolitan Greenspaces Master Plan, of which greenway planning is a part, is tied closely to this long-term plan and complements the urban growth boundary that contains growth in the metropolitan Portland area.

In Toronto, natural features and bioregions have motivated the greenway vision. The linkage of greenway planning to other watershed planning issues is becoming more widespread, aided by environmental legislation at the provincial level that complements greenway projects. In the past decade, awareness of urban watersheds has been elevated (via storm water management plans), but

Table 12.2. *Comparison of original greenway objectives*[a]

					Objective					
Greenway system	Recreation and fitness	Conservation and wildlife	Linking neighbour-hoods, parks	Education	Economic	Scenic quality	Non-motorised vehicular transport	Historic preservation	Growth manage-ment	Water quality protection
Chattanooga	•	•	•		•		•			•
Chicago	•	•	•	•		•	•			•
Minneapolis	•	•		•	•		•			
Portland	•	•	•	•				•	•	•
Toronto	•	•	•		•	•				•

[a] These objectives are abstracted from the planning documentation for each city, and do not necessarily represent ultimate outcomes and evolving goals. Objectives change over time, so this represents a snapshot in time when the plans were complete.

this awareness is not yet tied to smaller-scale greenway plans. Chattanooga's greenway network, especially its spine along the Tennessee River, has been strongly motivated by economic development goals and directly linked to various public–private partnerships for urban revitalisation.

Complementary policies and incentives, for instance in transportation, neighbourhood revitalisation, or biodiversity protection, may enhance greenway network implementation. For instance, a Chicago regional transportation plan focuses clearly on bicycle travel, with direct consequences for the development of non-motorised greenway routes. In addition, a statewide Forest Conservation Act demands that every development of 40 ha or more have a forest conservation plan, which could also provide linkages to greenway corridors.

The new frontier for urban greenway planning lies in multi-objective projects. Natural features protection and restoration will, like the restoration of neighbourhoods and communities, be central to successful implementation strategies for urban greenways. Searns (1995), in reference to this new generation of multi-objective greenways, writes

> Generation 3 greenway thinking addresses more complex issues and reflects more sophisticated thinking about environmental issues. Conveying these more complex needs in the economic and political decision-making arena will be one of the main challenges of the new generation of greenways.

12.4.2 Institutional structures

Each of the cases examined here is unique in its history, structure, and goals. A different set of stakeholders and agencies is involved in each city. Rarely, if ever, is the initiator of a greenways project the sole implementer. In order to realise projects of this size and complexity, a number of partnerships are inevitable and necessary at subsequent stages. These coalitions and partnerships are perhaps the strongest assets that predict completion of greenway corridors on the ground. The list of project co-operators is long and diverse for each city examined here. Some participants are active in only certain phases of projects; others are stable throughout the stages of project evolution.

At the federal level, the National Park Service Rivers, Trails and Conservation Assistance Program was heavily involved in three of the five projects. Their role as facilitators, networkers and brokers was critical. State government agencies often participate in greenway project development, through either planning or funding. (Several states, such as Maryland and Georgia, administer comprehensive greenway programmes at the state level.) In almost all cases, the Departments of Natural Resources and Transportation play key roles. Regional governments are typically involved, although they rarely exist in a form with extensive implementation power.

The role of the private sector is increasingly important. The co-operation of non-profit groups, foundations, and corporations helps give meaning and strength to community improvement, environmental protection and open space planning. The private sector also gives access to wealth. Non-profit groups are critical participants in all of the cases studied here. It is interesting that in some cases these groups start within government and are privatised as independent non-profit organisations in later stages. In Toronto, the Waterfront Regeneration Trust was established under provincial legislation to implement greenway objectives; it is now an independent non-profit organisation. In other cases, such as Chattanooga, the transition progressed in the opposite way. River City Company was originally a private non-profit group. Subsequently, its name was changed to River Valley Partners, and it became a joint public–private organisation with supplemental funding from the City to focus on a wide array of urban revitalisation issues. This highlights the diverse roles that government can have in both initiating and sustaining projects over time through co-operation with the private sector.

As shown in Figure 12.7, local and regional agencies are the primary implementers of greenways. But how are these agencies organised? Three patterns help explain their relationships. These patterns differ in level of control, authority and scope. In the first model, a regional entity carries out a strong role with primary control over implementation. It serves as an umbrella group over multiple jurisdictions. Portland is a useful example: its elected Metro Council has the power to implement greenway projects, primarily through land acquisition.

This institutional structure has several distinct advantages and disadvantages. A broad vision for the region can be co-ordinated across a diverse geographic area. Strong centralised leadership is capable of co-ordinating multiple agency efforts to accomplish land acquisition and development; more can be accomplished with a larger funding base from a wider geographic area. This approach allows more progressive jurisdictions to proceed without, or ahead of, slower, more reluctant jurisdictions. On the other hand, co-ordination efforts may break down when implementing the greenway plan is not a priority for each local jurisdiction. In addition, there may be less grassroots support and minimal public participation where a more top-down approach is used.

The second model uses a regional agency with moderate control to empower local jurisdictions through the initial planning stages and then partner with them for implementation and facilitates coordination between jurisdictions. Unlike Model 1 (Figure 12.7), neither the regional co-ordinator nor the local jurisdictions have primary control over funding. Three cases illustrate this

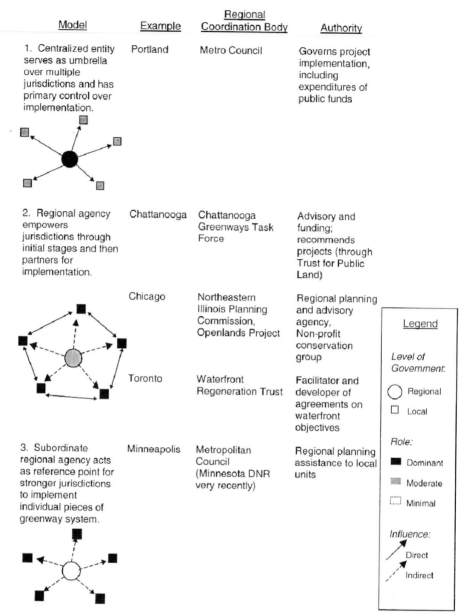

Model	Example	Regional Coordination Body	Authority
1. Centralized entity serves as umbrella over multiple jurisdictions and has primary control over implementation.	Portland	Metro Council	Governs project implementation, including expenditures of public funds
2. Regional agency empowers jurisdictions through initial stages and then partners for implementation.	Chattanooga	Chattanooga Greenways Task Force	Advisory and funding; recommends projects (through Trust for Public Land)
	Chicago	Northeastern Illinois Planning Commission, Openlands Project	Regional planning and advisory agency, Non-profit conservation group
	Toronto	Waterfront Regeneration Trust	Facilitator and developer of agreements on waterfront objectives
3. Subordinate regional agency acts as reference point for stronger jurisdictions to implement individual pieces of greenway system.	Minneapolis	Metropolitan Council (Minnesota DNR very recently)	Regional planning assistance to local units

Legend

Level of Government:
◯ Regional
☐ Local

Role:
■ Dominant
▨ Moderate
▢ Minimal

Influence:
↗ Direct
↗ Indirect

FIGURE 12.7.
Institutional models for implementation and acquisition of greenway projects.

model: Chattanooga, Chicago and Toronto rely on regional organisations for co-ordination, although in different ways.

In Chattanooga a national non-profit group, the Trust for Public Land, acts as a metropolitan co-ordinator. It organised the Chattanooga Greenways Task Force that serves as a co-ordination body for various local corridors.

Likewise, Chicago's Northeastern Illinois Planning Commission partners with non-profit organisations to facilitate greenway projects that are actually implemented by dozens of jurisdictions, forest preserves, and park districts. Similarly, Toronto's Waterfront Regeneration Trust is a facilitator and co-ordinator for local government units implementing the plan.

In the second model there is increased power at the local level, where leadership can develop among local jurisdictions. This may encourage co-ordination among jurisdictions. At the same time, the regional agency can prioritise funding from the state or federal level across the entire project. However, two main pitfalls may occur. Co-ordination between jurisdictions may be optional and implementation may be slower with shared responsibility across agencies.

In the final model (Figure 12.7), a subordinate regional agency acts as a reference point for stronger jurisdictions to implement individual pieces of the greenway system. Minneapolis/St Paul provides an example, where a regional board exists but has only recently been active in either planning or implementation of a regional greenways system, beyond recreational trails. Local units of government (cities and counties) plan and implement their own greenway projects that may or may not connect to their neighbours'. It should be noted, however, that the Twin Cities is moving towards Model 2, through the co-ordination provided by the state-run Metropolitan Greenways Program.

Strong local control is both the main advantage and disadvantage of this system. Grassroots efforts can take hold and projects can be accomplished at a manageable scale over a long period of time. There is ample room for private groups to put forth initiatives. However, there are definite drawbacks to this approach. Without regional leadership for co-ordination, a shared vision is lacking and regional connections are more difficult. In addition, there is less ability to take advantage of funding possibilities in a holistic way for the network.

12.5   Conclusions

Greenway implementation is neither simple nor linear. According to Mazmanian and Sabatier (1989) (in reference to implementing a broad range of environmental policies and programmes) '…there seems to be no routine or 'natural' progression to implementation. It may begin slowly or quickly, pick up or lose momentum, or pass through several cycles'. This uneven evolution describes several of the case studies in this research. Creating connected open space landscapes is inherently complex; the process may have neither a clear beginning nor an absolute end. Rather, the task is an evolving set of problems and opportunities that arise at different levels of decision-making over time.

Among the cases studied here, there is wide variation in the role of implementing organisations to guide this evolutionary process. Both centralised and decentralised schemes have advantages and disadvantages. Portland provides an example of a centralised natural areas and greenways system, where tangible successes have been achieved. At the same time, the Twin Cities' more decentralised method has also achieved marked success. For the Twin Cities, this may be due to a shared, but unstructured, priority within communities to forward the various objectives that greenways achieve. It is doubtful whether many other North American cities have achieved this level of commitment with a decentralised system. Over the twentieth century, this framework has been quite effective in the Twin Cities for protecting open spaces. However, connecting these lands on a larger scale will take new regional efforts.

The conventional wisdom is that governments are largely ineffective at long-range planning and implementation (Lowry 1998). Although the role of government is important in greenway development, especially in funding, it has its limitations. New governmental structures may not be needed, where existing agencies or non-governmental institutions might effectively take on the greenways charge. For example, the Northeast Illinois Planning Commission, in existence long before regional greenway planning began in the Chicago region, has been effective, especially in partnering with non-governmental facilitators. Chattanooga's example is useful as well; the scope of the regional greenways plan falls outside of any formal governmental structure to implement it. Ideas from the implementation literature describe a useful paradigm: '...the statement that a government is 'too large (or too small) to deal with a problem' often overlooks the possibility that the scale of the public and the political community need not coincide with that of the formal boundaries of a public organisation' (Ostrum *et al.* 1961). Informal arrangements between public organisations can perhaps achieve powerful changes in the physical structure of metropolitan regions.

Aside from government roles, scales, and processes, other factors predict open space protection in metropolitan areas. There is no simple way of understanding local land protection practices. Many factors interrelate to affect variation in open space protection: wealth, environmental attitudes, administrative capacity, residents' attachment to place, and social capital (see Press 1998, 1999). Even when a metropolitan greenways plan has been completed, political leaders are engaged, and implementation is underway these, and other, factors mitigate implementation efforts.

This research depicts some working models for greenways implementation. However, it leaves many unanswered questions. For Chicago, Chattanooga, Toronto, Portland and Minneapolis/St Paul, will the greenways vision last? Will each city's greenway network maintain steady progress toward implementation? Who will be involved? How will greenway implementation impact the community, in ecological, social and economic integrity? How will other North American cities begin to address the types of connectivity that greenways provide and what new models will be developed? It will be important to learn from the metropolitan regions that have pioneered greenway network implementation.

THOMAS S. HOCTOR, MARGARET H. CARR, PAUL D. ZWICK
AND DAVID S. MAEHR

13

The Florida Statewide Greenways Project: its realisation and political context

13.1 Towards a Florida Ecological Network

In the past 50 years, Florida has lost over 4.3 million hectares of natural communities including more than 50% of its wetlands and upland communities, including longleaf pine (*Pinus palustris*) forests and oak scrub that harbour much of the state's biological diversity (Myers 1990), and the loss of rural lands is continuing at a rate of over 50000 hectares per year (Matus 1999). Wide-ranging species such as the Florida panther (*Felis concolor coryi*) and Florida black bear (*Ursus americanus floridanus*) and other fragmentation-sensitive species have suffered serious declines, and Florida ranks third among states in the United States in the number of federally listed endangered and threatened species. However, efforts to protect Florida's natural heritage and biological diversity started to gain momentum in the 1970s with the beginning of an integrated land acquisition programme that expanded into a ten year, 3 billion dollar programme in 1990. Concurrently, conservation science evolved to support large-scale, integrated plans to protect biological diversity and other natural resources (Harris 1984; Noss and Cooperrider 1994; Soulé and Terborgh 1999a). Since the reports of Wright *et al.* (1933) and Wright and Thompson (1934) there has been a growing awareness that typical protected areas such as national parks are often not sufficient to protect viable populations of sensitive species and biological diversity as a whole (Noss and Harris 1986; Harris *et al.* 1996a, b). In the state of Florida the application of integrated reserve design principles has been promoted since the 1980s as a means to effectively conserve biological diversity in the face of rapid human population growth and habitat fragmentation (Harris 1984, 1985; Noss and Harris 1986; Noss 1987a; Harris and Gallagher 1989; Harris and Scheck 1991; Harris and Atkins 1991).

The Conservation Fund of Washington, D. C., and 1000 Friends of Florida began the Florida Greenways Program in 1991 with the goal of public endorsement and adoption of a greenways initiative. A governor-appointed commission was created, which completed a report in 1994, and greenways legislation was adopted the following year. The departments of Environmental Protection (DEP) and Transportation (DOT) funded the greenways plan through the federal transportation programme known as ISTEA (Intermodal Surface Transportation Efficiency Act). Although often misconstrued as simply a recreation programme, the greenways plan includes an ecological conservation component. In fact, considerable emphasis was placed on the definition, delineation, and implementation of the ecological component.

13.2   The Florida Ecological Greenways model

The University of Florida was hired to develop a decision support model to create a statewide greenways network. We applied Geographic Information System (GIS) technologies to assist in the physical design that included the identification of large areas of ecological significance and landscape linkages integrated as an interconnected statewide reserve network. Because Florida continues to experience rapid human population growth and habitat destruction (Harris and Atkins 1991; Kautz 1993), the identification and protection of a large-scale connected reserve network is essential to conserve the state's biological diversity. The goal of the analysis was to use a regional landscape approach to design an ecologically functional statewide greenways system that (Florida Greenways Commission 1994):

- conserves critical elements of Florida's native ecosystems and landscapes
- restores and maintains essential connectivity among diverse native ecological systems and processes
- facilitates the ability of these ecosystems and landscapes to function as dynamic systems
- maintains the evolutionary potential of the biota of these ecosystems and landscapes to adapt to future environmental changes

The rationale is that a functional reserve network integral to effectively conserving Florida's biological diversity and other natural resources can be identified by using a landscape approach guided by regional conservation planning principles combined with information on areas needed to protect viable populations of target species and natural communities. This was accomplished by incorporating assays of ecological significance, such as locations of rare and listed species, intact ecological communities, habitat areas needed to maintain viable populations of sensitive species, and land use data into a reserve design process integrating these components. The result is the first formulation of an interconnected Florida reserve system called the Florida Ecological Network (FEN), based on the definition by Forman (1995). This effort is similar to other projects in North America and Europe to identify, plan, and protect regional reserve networks (Jongman 1995; Soulé and Terborgh 1999).

13.3   Methods

13.3.1 Review process

Because of the complexity of the modelling process and the broad goals and objectives, we made numerous assumptions and decisions about specific modelling parameters and sequences. To ensure the appropriateness of our modelling decisions and to seek input on the use and application of data, we obtained technical input from 1995 through 1997 from the Florida Greenways Commission, the Florida Greenways Coordinating Council, scientists, university personnel, conservation groups, planners and other personnel in federal, state, and regional environmental agencies, and the general public in over 20 sessions.

Our GIS decision support model comprised four steps (Figure 13.1). The majority of the work was done in a raster format using ESRI's Grid module of the ArcInfo software package. The cell, or pixel, size for the analysis was 180 m × 180 m (approximately 3 ha). This size was used owing to the necessity to reduce data storage requirements and increase model simulation speed, and the requirements of analysis and mapping scales to remain within the USGS national mapping accuracy standards. Presently, the team is updating the results using a cell size of 30 m × 30 m because computer memory and speed have increased significantly.

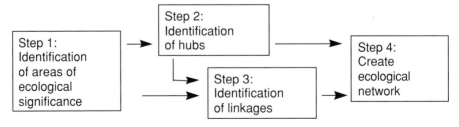

FIGURE 13.1.
The major steps in the Florida Ecological Network modelling process.

13.3.2 The modelling procedure

Step 1, the identification of areas of ecological significance, was performed through queries and reclassification of GIS data layers. These data layers included Strategic Habitat Conservation Areas, priority natural communities, existing conservation lands, roadless areas, and significant aquatic ecosystems. We developed rankings for each of the layers and combined all areas with primary ecological significance into a single layer called Priority Ecological Areas (PEAs) (Table 13.1). An area was included if it met one of the criteria, and no additional significance was given to areas that met multiple criteria. These areas represented the primary building blocks of the linked reserve network.

Step 2, selection of hubs, identified potential core areas for the protection of biological diversity and ecological processes. We began with the results of Step 1, then selected the areas with the highest ecological integrity potential. We accomplished this through the application of a five-step process.

1. We removed intensive land uses including improved pasture, croplands, residential, commercial, and industrial. This helped to rectify potential inconsistencies or errors in data used to determine areas of ecological priority. Land use data created by each of Florida's five water management districts (WMD) were based on 1988–94 satellite imagery and aerial photography. We also updated the most intensive land uses (urban) by using SPOT imagery from 1995–96 to ensure removal of areas that were no longer suitable.
2. We removed areas with extremely high road density (≥ 3 km km^{-2}) that greatly exceed general road density standards for protecting sensitive species (Noss 1992).
3. We removed areas with the greatest potential for negative edge effects, which were modelled coarsely as areas within 180 m of urban land uses. This was the minimum edge effect distance that could be modelled given our cell size and it was selected as a minimal estimate for the most intensive potential negative edge effects (Meffe *et al.* 1997).
4. We then identified remaining priority ecological areas that were 2000 contiguous hectares or larger, a figure based on recommendations from reviewers during model development. Such areas are large enough to support many species and ecological processes while also representing smaller areas of ecological significance (Forman 1995).
5. The resulting hubs were 'optimised' by smoothing edges and filling in internal gaps by adding lower-priority native habitat and potentially compatible land uses such as pine plantations and rangelands, which were identified by using a combination of the FWC land cover and WMD land use data.

Table 13.1. *Criteria for selecting Priority Ecological Areas for the Florida Ecological Network*

Data layer	Priority Area criterion	Explanation/rationale
FWC[a] Strategic Habitat Conservation Areas (SHCA)	All SHCAs	Includes lands outside existing protected areas needed to maintain or restore minimum viable populations of 30 focal vertebrate species, rare natural community types, important wetlands for wading birds, and globally rare plant species. Many focal species used in this analysis are umbrella species whose conservation requirements will meet the needs of other species and the natural communities identified represent a coarse filter approach to protect suites of species
FWC[a] Hotspots	Areas containing potential habitat for 7 or more focal species	Areas containing potential habitat for 7 or more of the focal species analysed in the identification of SHCAs. FWC staff recommended the threshold of 7
FWC[a] Wetland Hotspots	Areas containing potential habitat for 7 or more wetland-dependent species or 4 or more species requiring both wetland and upland habitat	Areas represent wetlands within Florida with habitat to potentially support additional wetland dependent and partially wetland dependent vertebrate species. FWC staff recommended the thresholds
FNAI[b] Areas of Conservation Interest (ACIs)	All ACIs	ACIs were identified outside existing public lands by using aerial photos, natural heritage data and expert knowledge. ACIs are high-quality, relatively pristine sites that contain occurrences of rare species
FNAI[b] Potential Natural Areas (PNAs)	All PNAs except those receiving the lowest rank owing to significant disturbance	Includes most of the remaining sites available to conserve native ecosystems in Florida, though some disturbance may be present and status of tracked species may not be completely known

Table 13.1. (*cont.*)

Data layer	Priority Area criterion	Explanation/rationale
Rare and priority natural community types based on FWC habitat data and rankings by Florida Natural Areas Inventory (FNAI)	All community types ranked S2 or higher that could be identified using the 22 class FWC land cover map that included coastal strand, dry prairie, sand pine and oak scrub, sandhill, tropical hardwood hammocks, freshwater marsh, and wet prairie	FNAI 'S' ranks are state ranks based on The Nature Conservancy's global rankings (G1–G5, 1 being most imperilled). The FWC land cover data are classified LANDSAT TM imagery from 1985 to 1989, but owing to the coarse scale of the classification, some S1 communities were not identified in this dataset. However, these communities were represented in the SHCA, ACI and PNA analyses
Existing public conservation lands and private preserves (e.g. Audubon, The Nature Conservancy)	All such lands	Approximately 20% of the state is now contained in conservation lands. Though management practices vary widely, all sites are potentially significant building blocks for a statewide reserve system
Proposed public conservation lands and easements	All such lands	Approximately 6% of the state has been identified for purchase through Florida's aggressive conservation land acquisition programme. These parcels were selected based on the presence of high-quality natural communities, habitat for rare species, opportunities to protect connectivity, or other conditions supportive of conservation objectives
Lands identified as part of the Coastal Barrier Resources Act	All such lands	These areas are typically coastal barrier islands identified by the federal government as undeveloped. Such sites are significant for conserving coastal ecosystems

(*cont.*)

Table 13.1. (*cont.*)

Data layer	Priority Area criterion	Explanation/rationale
Roadless areas	Areas 2000 ha or larger containing no roads of any kind	Roadless areas are important to species sensitive to humans, are typically buffered from disturbance and provide connectivity for species isolated by roads. A 2000 ha area was used based on federal roadless standards, average home range size for the Florida black bear (*Ursus americanus floridanus*), and recommendations by reviewers
'Roadless' areas without major roads	Areas 40 000 ha or larger containing no major roadways such as interstate, federal, or state highways, and large-capacity county roads	Large areas containing no high-volume roads may be critical for maintaining many sensitive species, especially wide-ranging animals such as the Florida black bear (*Ursus americanus floridanus*) and the Florida panther (*Felis concolor coryi*). The threshold is consistent with the FWC's objective to prevent major road construction in areas greater than 40 000 ha currently without major roads
State Aquatic Preserves, National Estuarine Research Reserves, Outstanding Florida Waters, Shellfish Harvesting Waters, Wild and Scenic Rivers	All such designated aquatic ecosystems	The greatest dearth in information about Florida's natural communities and species is in aquatic ecosystems. In the absence of such data, these designated aquatic areas, all indicating a level of quality that could support functional aquatic ecosystems, were used as a surrogate for a more comprehensive identification of significant aquatic features

Table 13.1. (*cont.*)

Data layer	Priority Area criterion	Explanation/rationale
Overlap Criteria	Moderately ranked FWC focal species, FWC wetland species hotspots, and lower ranked FNAI PNAs, smaller roadless areas (1000 ha or greater and 20000 ha or greater, respectively) that overlap with 100 year floodplains or areas of significant aquifer recharge	Moderately ranked habitat areas and roadless areas that overlap with areas significant for maintaining aquatic ecosystems and processes are also significant conservation features

[a] The Florida Fish and Wildlife Conservation Commission was previously named the Florida Game and Fresh Water Fish Commission.
[b] Florida Natural Areas Inventory.

In Step 3, we used the National Wetlands Inventory classification system (Cowardin *et al.* 1979) as the starting point for the derivation of three native landscape types: (1) coastal; (2) riverine and large wetland basins; and (3) upland-dominated (Table 13.2). The landscape types were used to partition hubs into the three general landscape classes for use in identifying linkages.

Next we identified five linkage types for hubs partitioned into the three landscape types: coastal to coastal, riverine to riverine, upland to upland, riverine to coastal, and cross-basin hub to hub. Linkages between hubs of like types were modelled before linkages between hubs of different types. The last linkage type, cross-basin or general hub to hub, was a broad category that allowed for exploration of linkage feasibility between dissimilar habitats or through agricultural landscapes where restoration may be needed to restore connectivity.

The algorithmic function 'least cost path' was used to identify landscape linkages. To apply the function, a suitability surface ('cost surface' in ArcInfo parlance) was created. Its purpose was to represent the relative suitability of every cell for potential inclusion in a linkage.

Five different suitability surfaces were created with one for each linkage type. In each suitability surface, the value assigned to each cell was inversely proportional to its relative suitability for that linkage type, (for example, a cell with the value of 1 is most suitable, 2 next, etc.) (see Figure 13.2, Table 13.3, Table 13.4). The function also allows for the identification of unsuitable cells where a

Table 13.2. *Landscape unit classification employed in landscape linkage identification for the Florida Ecological Network*

Landscape Unit	Ecosystems	

Coastal	open coastal waters
	coastal strand
	coastal salt marsh
	mangrove
		inshore marine habitats
		all other native habitats
		  within contiguous
		  100-year coastal flood
		  zone

Riverine and large
  wetland basins
	open waters of major Florida rivers (FREAC 1990), plus the
	  following when contiguous to major rivers or ≥400 ha:	

	bottomland hardwood
	  forest
	cypress swamp
	freshwater marsh and wet
	  prairie
	bay swamp
		mixed hardwood swamp
		shrub swamp
		freshwater aquatic habitats
		open lake waters

Upland-dominated	dry prairie
	flatwoods
	xeric scrub
	sandhill
	mixed hardwood–pine
	  forest
		hardwood forest
		tree plantations
		wetland/isolated aquatic
		  habitats when less than
		  400 ha

potential linkage could not be located. The relative suitability of each cell was determined by querying original data layers and data layers derived in steps 1 and 2. For 'like type' linkages, the suitability surfaces contained fewer criteria over a narrower value range than for dissimilar linkages. The least cost path function was then run to find the optimal path for selected hub pairs for each linkage type. Accepted paths were widened to include all contiguous cells of native habitat or lower intensity land uses up to a width of 25% of the linkage length to incorporate corridor widths that would provide wide landscape linkages, maintain habitat gradients from aquatic to upland ecosystems, and buffer aquatic ecosystems in riverine and coastal landscapes.

Step 4 combined the hubs and linkages to create the FEN.

13.4 Model results

The FEN links the larger public conservation lands while incorporating other important landscape features of each region (Figure 13.3). In the

Highly Suitable

Moderately Suitable

Unsuitable

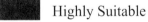

10   0   10   20   30 Kilometers

FIGURE 13.2.
Example from south-central Florida and the Kissimmee River basin (headwaters of the Everglades) of a suitability surface used for identifying potential riverine and large wetland basin landscape linkages and corridors. Original suitability values are aggregated in this figure into highly suitable, moderately suitable and unsuitable.

Table 13.3. *Riverine suitability surface values for the Florida Ecological Network*
The lower the value the higher the suitability.

Category	Value
Criteria for highly suitable areas	
Open water associated with major Florida rivers classified as PEA	1
Freshwater wetland ecosystems classified as PEA	1
Open water associated with major Florida rivers classified as SEA[a]	2
Freshwater wetland ecosystems classified as SEA	2
Criteria for moderately suitable areas	
Open water associated with major Florida rivers not classified as PEA or SEA	3
Freshwater wetland ecosystems not classified as PEA or SEA	3
Open water and areas with high road density or negative edge effect	4
Areas with high road density or negative edge effect that meet the riverine open water or freshwater wetland criteria for this linkage type	4
Criteria for unsuitable areas	
Intensive agriculture and urban lands	No value
All other cells	No value

[a] SEA or Significant Ecological Area, an area meeting criteria for moderate significance such as moderately ranked FWC hotspots and FNAI Potential Natural Areas, different form PEA.

Florida Panhandle the rivers that flow north to south form a network from the Blackwater River State Forest and Eglin Air Force Base to the Apalachicola National Forest. North-central Florida is dominated by the Suwannee River corridor, which links the lowlands of the Big Bend on the Gulf Coast to the Osceola National Forest – Pinhook Swamp – Okefenokee National Wildlife Refuge complex, and a large landscape linkage parallel to the western shore of the St Johns River that connects the Ocala and Osceola National Forests. In central Florida, river and swamp basins including the Kissimmee, Peace, St Johns, Myakka, and Withlacoochee Rivers join the sandhills and scrub of the Lake Wales Ridge and Brooksville Ridge and provide the primary elements of a network that includes the Ocala National Forest, Green Swamp, Three Lakes Wildlife Management Area, and Avon Park Bombing Range. Southern Florida is dominated by the Everglades National Park – Big Cypress National Preserve complex with linkages to central Florida via the Okaloacoochee Slough and Fisheating Creek, and via the Corbett Wildlife Management Area and the flatwoods, prairies, and sloughs east and north-east of Lake Okeechobee.

Of the approximately 9.3 million hectares (57.5% of the state) incorporated into the FEN, 4.8 million hectares (52.2% of the network) are within existing public and private conservation lands, or open water (considered public

Table 13.4. *Hub to hub suitability surface values for the Florida Ecological Network*
The lower the value the higher the suitability.

Category	Value
PEAs that meet all but the 2000 ha size criteria for hubs and are contiguous with significant coastal and/or inland aquatic features	1
Other PEAs that meet the 2000 ha size filter	2
SEAs[a] that are contiguous with significant coastal and/or inland aquatic features	2
Native habitat that is contiguous with significant coastal and/or inland aquatic features	3
All remaining SEAs	3
All other native habitat	3
Low-intensity land use or land cover that is contiguous with significant coastal and/or inland aquatic features	4
All other low-intensity land use or land cover	5
Native habitat lands with areas of negative edge effects or areas of high road density	600
Lands with low-intensity use and areas of negative edge effects or areas of high road density	700
Improved pasture contiguous with significant coastal and/or inland aquatic features	7000
Cropland contiguous with significant coastal and/or inland aquatic features	8000
All other lands in moderate-intensity use, contiguous with significant coastal and/or inland aquatic features	9000
Improved pasture	70 000
Cropland	80 000
All other lands with moderate-intensity use	90 000
Open water	100 000
Urban lands	No value
All other cells	No value

[a] SEA or Significant Ecological Area, an area meeting criteria for moderate significance such as moderately ranked FWC hotspots and FNAI Potential Natural Areas.

domain by Florida law). Thus, less than half of the identified FEN is on private land that may need protection (Table 13.5). Of the private land included in the FEN, more than 50% is within an existing conservation project, wetlands, or 100-year floodplains.

In reserve design, representation analysis is conducted by comparing the occurrence of features of ecological significance, such as natural communities or focal species, with the occurrence of protected areas to determine what features

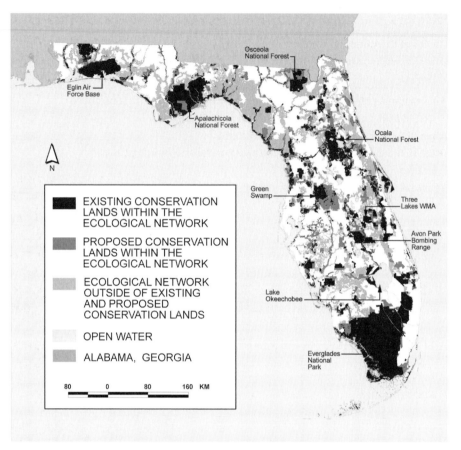

FIGURE 13.3.
Florida Ecological Network model results including existing and proposed
conservation lands within the ecological network. Existing conservation lands
include all public lands with at least some conservation management and private
preserves such as lands managed by The Nature Conservancy. Proposed
conservation lands include all projects within official federal, state, regional and
local land acquisition or protection programmes. Although the ecological network
includes water within all of the major rivers and most intact estuarine systems,
these features have not been differentiated from other areas of open water in
this figure.

require greater protection (Scott *et al.* 1993; Noss 1996). A coarse representation
analysis was done for the FEN by comparing total existing hectares of major
ecological communities (Cox *et al.* 1994) with the amount found within existing conservation areas and the FEN. It is clear that the identified FEN significantly enhances the protection of each community type (Table 13.5). The
percentage increases in protection for sand pine scrub, xeric oak scrub, and
longleaf pine sandhill are of particular significance. These community types
are endangered globally and provide habitat for many endangered, threatened,

Table 13.5. *Comparison of the total land area of existing ecological community types (habitats) in the state of Florida with area of habitat types found in existing conservation lands and the amount included in the Ecological Network*

Ecological community type	Total area (ha)	Area in existing conservation lands (ha)	Total habitat in existing conservation lands (%)	Area in Ecological Network (ha)	Habitat in Ecological Network (%)	Increase in protected area (ha)	Percentage increase in protected area
Coastal strand	4281	3145	73.4	3475	81.2	330	7.7
Dry prairie	519895	133334	25.6	422050	81.2	288716	55.5
Pinelands	1651235	413066	25.0	1076578	65.2	663512	40.2
Sand pine scrub	131708	105501	80.1	117500	89.2	11999	9.1
Sandhill	390056	146250	37.5	248888	63.8	102638	26.3
Xeric oak scrub	52628	22272	42.3	38967	74.0	16695	31.7
Mixed hardwood–pine forests	387889	46532	12.0	197300	50.9	150767	38.9
Hardwood forest	818030	185651	22.7	530194	64.8	344544	42.1
Tropical hardwood forest	5872	3313	56.4	4210	71.7	897	15.3
Saltmarsh	195710	121584	62.1	182616	93.3	61032	31.2
Freshwater marsh/wet prairie	1153285	778923	67.5	1023724	88.8	244801	21.2
Cypress	621504	253454	40.8	546964	88.0	293509	47.2
Mixed hardwood swamp	1076484	276915	25.7	864698	80.3	587783	54.6
Bay swamp	57968	16763	28.9	47102	81.3	30339	52.3
Shrub swamp	252327	152976	60.6	222950	88.4	69974	27.7
Mangrove swamp	229012	198100	86.5	221703	96.8	23603	10.3
Bottomland hardwoods	40033	23532	58.8	39926	99.7	16394	41.0

and endemic species as well as candidate species for listing as threatened and endangered (Myers 1990; Noss *et al*. 1995; Platt 1998).

An additional form of representation analysis was conducted by comparing the FEN results to information for natural communities and rare species from the Florida Natural Areas Inventory (FNAI). Of 69 natural communities contained in the FNAI data, all had at least one occurrence in the FEN, and only four had less than 50% of their occurrences within the FEN. The analysis of rare species occurrences indicates there are 32 species (6% and mostly plants) not known from existing conservation lands or within the FEN. Therefore, most rare natural communities and species are represented in existing conservation areas and the FEN, but more work needs to be done to enhance the accuracy and completeness of the analysis.

The FEN identified as part of the Florida Statewide Greenways planning process is a significant step towards protection of an integrated state reserve system. In the 1980s Harris (1985), Noss (1987a), and The Nature Conservancy (Noss and Cooperrider 1994) recommended linked reserve systems through intuitive representations of networks and mapping charrettes. Subsequently, the Strategic Habitat Conservation Area analysis by the Florida Fish and Wildlife Conservation Commission and the natural areas identification by the Florida Natural Areas Inventory provided systematic assays identifying priority areas for conservation (Mann 1995). But the progress represented by the design and execution of the FEN delineation process was the combination of a systematic landscape analysis of ecological significance and the identification of critical landscape linkages in a way that can be replicated, enhanced with new data, and applied at different scales. The FEN connects and integrates existing conservation areas and unprotected areas of high ecological significance. This information can be used in concert with other information on conservation priorities to develop a more integrated landscape protection strategy. Such an integrated reserve network will more thoroughly protect important ecological functions, community and landscape juxtapositions, and the need for biotic movement than the present collection of isolated conservation areas (Noss and Harris 1986; Harris *et al*. 1996a).

13.5 Discussion

Each of us is an artist whose task it is to shape life into some semblance of the pattern he dreams about. The molding is not of the self alone but of shared tomorrows and times we shall never see. So let us be about our task. The materials are very precious and perishable.

(*Arthur Graham, on a plaque in the Chapel of Warren Wilson College, Swannanoa, North Carolina*)

13.5.1 Prioritisation of the Florida Ecological Network

Since such a large portion of the state was identified as being potentially suitable as part of statewide ecological network, prioritisation is important to facilitate the use of the FEN as a planning tool. We are currently completing a prioritisation effort for the FDEP to identify the areas that represent the best opportunities to maintain large, connected landscapes that will best conserve biological diversity over the long term and maintain essential ecological processes and services including water quality and quantity protection, protection from storms, clean air, nature recreation, etc. The prioritisation includes two major components. First, priorities based on ecological significance including the importance of ecological resources and the potential for functional connectivity are identified. Second, a model assessing development pressure is being created to identify areas most threatened by conversion to intensive land uses. These two components will then be combined to identify critical landscapes and ecological linkages that are the highest priorities for conservation action in the immediate future.

13.5.2 Potential applications

The FEN is a representation of areas most suitable for inclusion in a connected, statewide reserve system. However, this representation is also a powerful planning tool that can be used in a variety of applications including land acquisition, development of a statewide conservation plan, road corridor selection, and ecological mitigation. Since Florida is continuing with an ambitious land acquisition programme, discussed in more detail below, the FEN can be used to make land acquisition planning more integrated and comprehensive. In the past, land protection has often focused primarily on the intrinsic characteristics of particular sites and not their potential contribution to a larger, integrated system. The FEN fills this gap by elucidating the contextual significance of all sites within a statewide network of conservation areas.

The FEN is also a key part of a potential statewide conservation plan. Such a plan would identify all the sites needed to protect Florida's natural heritage and natural resources. Though the FEN encompasses the vast majority of the land that might be included in such a plan, not all important sites and species are contained within the model results, including some of the remaining globally imperilled pine rocklands in south-east Florida (Snyder *et al.* 1990) and oak scrub tracts along the Lake Wales Ridge that support many rare species (Myers 1990).

As mentioned above, the Florida Department of Transportation funded the majority of the modelling effort. Their interest was primarily to improve transportation corridor selection and ecological mitigation. A burgeoning human population has created political pressure to continually build new roads and expand existing roads, with an average of 3 km of new highway built in Florida per day over the last 50 years (Harris and Gallagher 1989). This transportation system has a variety of impacts on Florida's ecosystems and biodiversity. Impacts include roadkills of various sensitive species including the Florida panther and Florida black bear (see Chapter 5). The FEN will be used to minimise future impacts by strategically placing new and expanding transportation facilities. When transportation and other development projects have ecological impacts, the FEN can help locate mitigation sites that will contribute to a larger reserve system.

In rapidly growing segments of Florida, new development threatens to eliminate opportunities for maintaining or restoring ecological connectivity. In south Florida, either the South Florida Water Management District or the US Army Corps of Engineers must permit the majority of large proposed developments. Consultation with state and federal wildlife agencies provides the opportunity to alter development design and (rarely) to deny development permits.

More often, permit applications are approved after an acceptable amount of mitigation has been offered to 'offset' development impacts. In reality, traditional mitigation is nothing more than strategic withdrawal whereby an existing amount of habitat is purchased elsewhere to replace that which is displaced by development (Maehr 1997). Thus, the typical result is a net loss of habitat and nothing is truly mitigated. Effective conservation for species such as the federally endangered Florida panther could be facilitated by identifying strategic node, network and corridor habitat and requiring off-site mitigation to occur in lands identified by the FEN. Private sector initiatives such as mitigation banks, if established within the FEN, could serve as significant hedges against future development, and would reduce the need to rely on public funds to acquire such lands. Typically, mitigation lands are either turned over to a government land management agency, or set aside in perpetual conservation easements. The most effective mitigation would occur when degraded habitat was restored as a functional part of the FEN. Such a practical application of the model results would elevate its value as a biodiversity conserving tool, clarify a regulatory process that is often muddied by ambiguous and inconsistent enforcement, and offer a level of co-ordination that has not existed in the past.

Although we believe the FEN is a significant step forward in Florida's conservation strategy, there are still many issues and questions that need to be addressed. Protection of connected, large-scale conservation reserves may require many more years for implementation, but the future of Florida's natural heritage, biodiversity, natural resources, and ecological integrity and services is in doubt without such a comprehensive approach. Below, we outline and discuss the major impediments to implementing a FEN and the factors and trends that will help facilitate protection.

13.5.3 Potential cost and the political and cultural climate

With approximately four million hectares of private land in the FEN, the cost of protection through fee-simple acquisition, conservation easements, and/or agreements with landowners will be high and the period of implementation will be long. Through the land acquisition programme called Preservation 2000, Florida committed approximately US$3 billion to protect over 500000 ha. Fortunately, Florida has continued a strong commitment to land protection through legislative passage of a successor programme called Florida Forever. Although this programme is also funded for US$3 billion, fewer dollars will go towards land protection than in the Preservation 2000 effort, as some of the funding is to be spent in urban areas and on water-related projects. So with a rough estimate of approximately US$2 billion for land protection per decade from the state, plus some additional support from federal land acquisition programme, we estimate it will take 3–4 decades to protect most of the private land identified in the FEN.

Although the people of Florida have demonstrated a willingness to protect Florida's natural resources through aggressive land acquisition, politics and future uncertainties may significantly hinder the protection of the FEN. Since its inception the United States has often been dominated by a 'frontier' mentality that has led to policies promoting the privatisation of land and other natural resources in an attempt to maximise economic development (Sax 1991). Although the United States has also developed strong conservation and environmental policies since the 'closing of the frontier' in the late nineteenth century, a strong commitment to private property rights and the privatisation of resources has remained. In the past two decades, private property rights groups have increased in visibility and political activism. The growth of such groups has been fuelled by private corporations in the mining, timber, construction, and other industries that have a large stake in the continuing exploitation of natural resources, and through a dislike of what is perceived as 'Big Government' and distrust of government motives (Noss and Cooperrider 1994). The 2001 appointment of Gale A. Norton as Secretary of Interior in the Bush Administration and similar industry insiders to other environmental positions is witness to the power of these entities and their determination to influence government policy (Jehl 2001; Seelye 2001). In Florida, private property rights groups and supportive politicians have also criticised land acquisition as government land grabs and argue that such conservation efforts reduce tax revenues and stymie economic development. Although such arguments can be generally dismissed as short-sighted, private property rights as defined by the Fifth Amendment of the Constitution are an integral part of American law and politics and will continue to influence land use decisions and other policies for environmental protection.

Conservation efforts in Florida also are hindered by a state economy that is increasingly dependent on construction and development associated with a burgeoning human population. Florida's population is currently almost 16 million people, and approximately 300000 residents are added each year (US Census Bureau 2000, 2001). Rapid population growth has created an economy that drives politics and government policy. It has been estimated that at least 1 million of the state's 4.7 million jobs can be directly and indirectly attributed to the construction industry (Nicolas and Steiner 2000). Such a political climate has created a political and economic system supportive of rapid residential and commercial development, including the development of large-scale, intensive road networks. These policies and processes have led to rapid habitat loss and fragmentation, with estimates of annual habitat and rural land loss ranging from 50000 ha (Matus 1999) to as high as 60000 ha per year (Harris and Scheck 1991; Harris and Silva-Lopez 1992). These losses are driven by an economy and political system that often considers land as merely a commodity used to maximise economic profit. Unfortunately, the viewpoint of Peter Rummel, the CEO of the St Joe Company, Florida's largest private landowner with over 400000 hectares, typifies this common attitude: 'The thing that goes in one end is a piece of land. The thing that comes out the other end is a home or an office or a resort hotel or a golf course...' (*Word*, July 4, 1999).

Florida enacted growth management legislation in the 1970s and 1980s in response to rapid urbanisation and suburbanisation that created state oversight for local future land use plans and large developments with regional impacts. Drafters of the legislation felt that state oversight would expose the link between growth and infrastructure costs and as a consequence might slow growth to a pace that could be met by infrastructure improvements and direct growth to areas most suitable (Nicolas and Steiner 2000). The effectiveness of the legislation, however, has been questionable; for example, it is estimated that if population growth continues at its current rate, the 2099 population would reach 455 million people (Nicolas and Steiner 2000), a figure certain to be beyond the capacity of the state's water resources, if not land resources. This lack of success is acknowledged by nearly everyone in Florida. The February 2001 Florida Growth Management Study Commission's Report states that 'the quality of growth has not met our expectations'. In this same report key recommendations include: 'empower citizens to better understand

and participate in the growth management process' and 'establish a primary vision statement for Florida with a healthy, vibrant and sustainable economy as its priority'. These recommendations represent the tension between state-mandated, top-down planning versus locally driven, bottom-up decision making. Opponents of bottom-up control contend that this will only further open the pro-development floodgates as those who profit by maximising development and not environmental protection often dominate local politics. Nathaniel Reed (2000), Assistant Secretary of the Interior under Richard Nixon, and founder of a growth management advocacy group called 1000 Friends of Florida, puts it like this: 'As long as there are county commissioners who are more interested in helping developers than they are in protecting the environment, we're not going to get anywhere'.

Though some local governments in Florida have developed strong conservation policies including local land acquisition programmes, such efforts frequently occur in counties that have already been largely developed and where opportunities to protect isolated sites of high ecological integrity, much less connected reserve systems, have disappeared. In contrast, counties in northern Florida and elsewhere that are still predominantly rural generally encourage development despite the fact that similar policies have led to massive urban development in other parts of the state.

The 'top-down' versus 'bottom-up' debate is of further relevance to the FEN specially and landscape ecology in general. Landscape ecology analyses require a look at the next larger context (Forman 1995). Yet the current trend toward local decision-making inherently resists 'mandates' from above that suggest or recommend policies. The FEN is clearly such a 'mandate' and is likely to be resisted to some degree at the local level. Implementation is still highly probable, however, because of the presence of the state land protection programme, Florida Forever, and environmental activists who have learned how to be effective at the local level.

It has become clear that effective conservation of biodiversity and other natural resources will require a large-scale, integrated and comprehensive approach that does not tread on private property rights. Integral to such an approach will be a combination of local participation and decision making that can build a conservation constituency with top-down planning and oversight that comprehensively identifies ecological resources of statewide significance and promotes land use decisions that both meet local needs and protect the state's natural heritage and ecological services. We agree with Meffe and Carroll (1997) that: 'In order for participatory decision making to contribute to ecosystem sustainability, mechanisms must be found to balance self-interest with the overriding need for resource sustainability.'

Table 13.6. *Comparison of the Florida Ecological Network and other ecological resource inventories and existing and proposed conservation lands*

Comparison categories	Area of category within model results (ha)	Percentage of model results	Percentage of category within the model results	Area of category in study area (ha)	Percentage of state
State	N/A	N/A	N/A	16175928	100.0
Ecological Network model results	9298742	100.0	N/A	9298742	57.5
FWC[a] Strategic Habitat Conservation Areas (SHCAs)	1586567	17.1	80.6	1968587	12.2
FNA[b] Areas of Conservation Interest and Potential Natural Areas (ACIs)	1521085	16.4	68.7	2214813	13.7
Existing or proposed conservation lands; open water; SHCAs; or ACIs	7539052	81.1	81.4	9259270	57.2

[a] Florida Game and Fresh Water Fish Commission (renamed Florida Fish and Wildlife Conservation Commission in 1999).
[b] Florida Natural Areas Inventory.

13.5.4 Debate about connectivity

The need for setting conservation priorities also involves the debate about the importance of protecting corridors *versus* protecting core areas of high-quality habitat (Simberloff and Cox 1987; Noss 1987b; Simberloff *et al.* 1992; Hobbs 1992; Beier and Noss 1998). How are decision-makers to choose between these alternative strategies? The Florida Greenways Project provides some insight. First, these approaches are not mutually exclusive; prioritisation of land protection can include both a larger connected network and isolated sites. However, landscape linkages that contain high-quality habitats needed to maintain viable populations of sensitive species can also be identified. The high degree of overlap between the FEN, Strategic Habitat Conservation Areas, and priority sites identified by the Florida Natural Areas Inventory suggests that this will occur frequently (Table 13.6).

Next, landscape linkages needed for facilitating connectivity for wide-ranging species should also be identified as priorities. Although the FEN model identified landscape linkages that may provide functional connectivity and promote the re-establishment of wide-ranging species, specific analyses should be conducted for both the Florida panther and Florida black bear to further explore this possibility. Broad landscape analyses of connectedness are useful, but species-specific analyses are essential for determining potential for connectivity of particular populations and identifying minimal viable areas for metapopulations (Beier 1995, 1996; Maehr and Cox 1995; Beier and Noss 1998). Although there is likely to be additional debate about landscape linkage and corridor projects, connectivity has been accepted as a critical reserve design principle (Harris *et al*. 1996b; Beier and Noss 1998). Because natural landscapes are generally connected, the burden of proof should be on those that remain sceptical about the need to protect landscape linkages and corridors and not vice versa (Noss 1987b, 1991; Beier 1996; Beier and Noss 1998). In human-dominated landscapes there will often be many more opportunities to protect isolated sites than larger connected landscapes, but the opportunity to protect existing landscape linkages or to restore them will diminish rapidly as the human population continues to grow. As stated by E. O. Wilson (2000), 'Now is the time to create systems of reserves, because the window of opportunity is closing fast.'

Third, isolated sites of ecological significance can be identified and also prioritised. Isolated sites can contain critical elements of biodiversity (Shafer 1995). Such sites can provide valuable habitat for less mobile species or species capable of using isolated sites as stepping stones. In Florida, isolated patches of globally imperilled pine rockland (Snyder *et al*. 1990) and scrub communities (Myers 1990) contain many endemic species that can be protected nowhere else.

13.5.5 Interagency co-ordination

An ambitious conservation strategy like the FEN will require significant interagency co-ordination. The experiences encountered during the development of the FEN provided reasons for both hope and pessimism. The greatest obstacles to co-ordination seemed to occur when the roles of the agencies were not clearly defined and when funding for an agency was not secure. In the Florida case, it was interesting that the greatest challenges in this regard were found within the lead state agency (DEP).

This experience is common among government agencies. In a classic book written about government function, Seidman (1975) called interagency committees 'the crabgrass in the garden of government'. But as noted by Lambright (1997), 'efforts to extirpate them seldom succeed…because alternatives generally are worse'.

In evaluation of a federal interagency committee Lambright (1997) points out that success is possible when there are (1) common interests, (2) supporting constituencies, (3) good morale among participants and (4) strong leadership. The Florida Greenways Project experience was consistent with this observation. The most effective committee participation came from those agencies possessing these four traits. Federal participation in the Florida project was particularly laudable, but local and regional agencies (water management districts) also participated constructively.

It was also true that a clear and methodical process significantly contributed to the success of the interagency co-ordination. The purpose of each meeting was clearly defined and progress from one meeting to the next was guaranteed. The result was that time spent in meetings led to consensus building, and now that the FEN is in the implementation stage, that consensus is contributing to greater success. The time dedicated in early project phases pays off in the final critical stage.

13.5.6 Federal, state and local actions and the nationwide trend for greenspace protection

Implementation of the FEN will benefit from the actions occurring at the federal, state and local levels and by the broad and bipartisan support for greenspace protection. On 29 November 1999 President Clinton signed an omnibus spending bill that appropriated US$464 million in fiscal 2000 for land acquisition. Some of that was specifically earmarked for Everglades acquisition and restoration. In 2000 it appeared the Conservation and Reinvestment Act (CARA) might pass through Congress, and support from Clinton was virtually guaranteed. However, in the flurry of pre-election activity the bill became stymied in late 2000. It did not die, however, and in 2001 there was considerable optimism that it would pass into law before the end of the year. In its current state, the bill would provide US$1 billion annually for marine conservation and US$900 million for the Land and Water Conservation Fund. There appears to be broad, bipartisan support for the Act including support from the President Bush and Secretary of the Interior Norton (Associated Press, June 2001). This kind of federal funding for land acquisition will nicely complement the state activity.

Florida now has 27 counties and municipalities with local land acquisition programmes. In 2000, voters approved the initial adoption or reauthorisation of six local programmes (Reed 2000). During the 2000 election cycle, no proposed local programme was defeated. As with Florida Forever, these local programmes have good potential to assist in implementation of the FEN. There will likely be a tendency for local monies to be spent on smaller and more urban sites, yet it has already been proven in places like Volusia and Brevard Counties that some funds will be directed towards large sites and parcels contributing to connectivity.

Nationally there is clearly broad support for greenspace protection. Some say Florida has helped set the national trend with its Environmentally Endangered Lands Program of the 1980s, Preservation 2000 Program of the 1990s and the

BOX 13.1. The Florida Forever Program

In Florida the adopted measures have been initially applied in the first round of project selection. The result was the inclusion of several large projects and several projects that contribute to ecological connectivity. This is a very encouraging development as it represents the institutionalisation of core concepts embodied by the FEN.

Florida Forever, a 10-year land acquisition programme was adopted by the Florida legislature in 1999. Implementation began in 2000 with the creation of the Florida Forever Advisory Committee, charged to develop measures to be used in acquisition project selection. Our active participation in the development of the measures helped result in the adoption, under the broad goal of biodiversity protection, of Measures B3 and B5, which are especially relevant to FEN implementation.

GOAL B: INCREASE THE PROTECTION OF FLORIDA'S BIODIVERSITY AT THE SPECIES, NATURAL COMMUNITY, AND LANDSCAPE LEVELS:

Measure B1:   Acres acquired of significant Strategic Habitat Conservation Areas (SHCA).

Measure B2:   Acres acquired of highest priority conservation areas for Florida's rarest species.

Measure B3:   Acres acquired of significant landscapes, landscape linkages, and conservation corridors, giving priority to completing linkages.

Measure B4:   Acres acquired of under-represented native ecosystems.

Measure B5:   Number of landscape-sized protection areas that exhibit a mosaic of predominantly intact or restorable natural communities (>50000 acres) established through new acquisition projects, or augmentations to previous projects.

Measure B6:   Percentage increase in the number of occurrences of endangered/threatened/special concern species on publicly managed conservation areas.

new Florida Forever Program (see Box 13.1), but the momentum generated by Florida is being amplified. In 1998, 148 state and local open space protection referenda were on the November ballot; 124 (84%) of these passed and 24 or (16%) of them failed (Land Trust Alliance 1999). This trend continued in 2000 when 90% percent of 102 referenda passed, authorising more than US$1.8 billion in local taxing authority and bonds for open space protection (Land Trust Alliance 2001). This trend will support comprehensive conservation strategies like the FEN, around the country.

13.5.7 Private sector initiatives

Increasingly land conservation is being pursued through private sector initiatives. These initiatives provide an important complement to public initiatives and will enhance the implementation of a comprehensive ecological network. Among the specific examples in Florida is Florida's Legacy, an organisation preserving Florida's unique natural heritage through co-operative private initiatives and public bio-education. Florida's Legacy is the manager of the 12 000 ha Mallory Swamp Restoration Project. Landowners partner to set goals and fund the conservation, restoration, and education activities of Mallory Swamp. Another example of private sector initiatives is the growth in the land trust movement. The Red Hills Conservation Program that is part of the Tall Timbers Research Station, which works in the area between Tallahassee, Florida and Thomasville, Georgia, has proven to be particularly successful. Currently they have approximately 24 000 ha under conservation easement (Tall Timbers Research Station, www.talltimbers.org, May 2000).

13.5.8 Increasing use and utility of conservation easements

Land protection programmes in Florida continue to shift toward greater emphasis on the use of conservation easements and agreements. The Florida Department of Environmental Protection adopted an ecosystem management approach that includes conservation agreements as a critical component (FDEP 1998), and the new land acquisition programme, Florida Forever, includes a much greater emphasis on easements than past initiatives. There are disadvantages to this approach (such as monitoring compliance), but the benefits include significantly lowering land protection and management costs (Daniels and Bowers 1997). Although there will still be a need to acquire and protect high-quality habitat for the most sensitive species, lands supporting lower-intensity land uses can provide habitat, potential linkages, and buffers for wide-ranging and other species of conservation interest (Harris 1984; Noss and

Cooperrider 1994). Conservation easements can maintain such lands in current uses while establishing management practices that are more compatible with conserving biological diversity. Such lands are also maintained on local property tax rolls.

13.5.9 Road impacts, issues, and opportunities

Another challenge is to retrofit the existing highway system in Florida and plan future road projects to be as compatible as possible with the protection of a statewide reserve system (see chapter 5). The Florida Department of Transportation has made significant progress including the construction of a comprehensive system of underpasses where Interstate 75 crosses the Big Cypress National Preserve to allow Florida panthers and many other species to cross safely under the highway (Foster and Humphrey 1995). One underpass has been constructed at a black bear roadkill hotspot in central Florida and more are planned (Roof and Wooding 1996). A comprehensive assessment of all potential interfaces between major roads and priority ecological conservation areas for future mitigation (for example, lengthening existing bridges and culverts, constructing new wildlife underpasses) co-ordinated with the FEN modelling process, has also recently been completed (Smith 1999a).

However, there is still a tendency to avoid the construction of adequate underpass facilities in many places where they are needed, and the construction of crossing structures for wildlife still has to compete with what are considered to be higher priorities for transportation infrastructure funding. Also, although some wildlife underpasses (such as the I-75 structures) have proven to be successful in some situations for certain species, the efficacy of relatively narrow crossing structures on limited stretches of highway for mitigating fragmentation and roadkill impacts is equivocal or even unlikely. Some species, such as the Florida black bear, sometimes avoid underpass structures or climb fences meant to funnel them towards crossing structures and keep them off roadways (Roof and Wooding 1996). More comprehensive mitigation of existing road impacts may be needed, including the elevation of large swathes of highways that cross existing conservation lands or natural movement corridors such as rivers and other riparian areas. Impacts from building new highways should be avoided, and the FEN can be used as a primary planning tool for avoiding new road projects that fragment intact landscapes or wildlife corridors.

Florida and other states also need to take advantage of federal transportation enhancement money that can be used for environmental mitigation and transportation improvements that increase the compatibility of transportation infrastructure with conservation objectives, including the construction of wildlife crossing structures. The federal Transportation Equity Act for the 21st Century (TEA-21) allocates at least 10% of Surface Transportation Project funds for environmental enhancement projects through the year 2003. In 2001, this funding source provides at least 25 million US dollars of funding for environmental enhancements in Florida and over 545 million nationally (http://www.fhwa.dot.gov/tea21/index.htm).

13.6 Conclusions

Reserve design is an iterative process that must continually consider new information. Work on refining and enhancing the FEN is progressing through several projects and scales. We are continuing efforts to prioritise elements within the FEN including co-ordination with additional species-specific habitat analyses. Florida's federal GAP analysis project has yet to be completed, the Florida Fish and Wildlife Conservation Commission continues to analyse additional species, and The Nature Conservancy is engaged in ecoregional planning including many focal species and all natural communities. These projects likely will identify priorities to be addressed in future iterations of a state reserve system plan, and, as always, field assessments of priority sites need to be done as part of the protection process. As land development continues, loss of habitat will have to be monitored and conservation plans adjusted as necessary. Also, considering Forman's comments on the 'ethics of isolation' (1987), we are working with Region IV of the US Environmental Protection Agency to identify a regional ecological conservation network for the south-eastern United States, which could lead to coordination with other efforts to identify and protect reserve networks in North America (Soulé and Terborgh 1999).

The Florida Forever acquisition programme provides considerable reason for optimism. Not only was the programme reasonably funded by the legislature, the incorporation of the measures of connectivity and size into evaluation criteria was a triumph. Additionally in 2001, there was legislative action for rural lands protection that might augment Florida Forever (Florida Audubon Society 2001). Based on one of the eight recommendations of the Growth Management Study Commission, 'Develop an incentive based state rural policy which restores rural land values and protects private property rights, including dedication of additional revenue for public purchase of conservation and agricultural easements and a special overlay of transferable density allocations for rural property to be used for the implementation of cluster development in appropriate locations,' the Rural and Family Lands Protection Act was passed. Included is a rural stewardship programme that will support five local governments in the initiation of transfer of development rights pilot projects to address urban sprawl and the potential use of conservation easements to protect rural lands. Although the mandate remained unfunded in the 2001 legislative session, full funding will be a high priority for conservation organisations in 2002.

The purchase of lands in Florida through Save Our Rivers funding at the water management district level continues apace. Even though the water management districts must place water resource protection as their highest priority in evaluating acquisition opportunities, this goal is increasingly leading to purchases that also address connectivity. Notable examples of this are the Suwannee River Water Management District's determination to buy lands along the Suwannee River, a major north–south riparian system, and the similar actions of the St Johns River Water Management District, especially for the upper St Johns River.

Regardless of the level of conservation activity, federal, state, regional or local, open space protection seems to be enjoying broad and bipartisan support. This is cause for considerable encouragement as lawmakers are so evenly split on most other issues. Issues that entertain support from both sides of the aisle are likely to be those that move through Congress and state houses most rapidly.

The greatest threat to the successful implementation of the FEN is Florida's rapid population increase, mostly from in-migration. The race is on between the conservation of lands and their development. Conservation is challenged by the pace and spatial distribution of development, which results in habitat loss and fragmentation of the habitat that remains. On the positive side, conservation has a running start with over 25% of the state already protected. On the negative side, the booming economy, low tax structure and ironically, the beautiful environment fuel development and concomitant habitat loss and fragmentation. The FEN was developed as a planning tool to assist in strategic selection of conservation expenditures and it appears it is being used for that purpose. Regardless, however, it is not yet clear how the race will end.

Perhaps as soon as 2050, Florida's lands will be permanently allocated to one of three uses: conservation, agriculture or human settlement. Beyond that point, there may be some tinkering with density here or modest conversion of agricultural lands there, but the overall pattern of land use will be decided. The elaboration of the Florida Ecological Network is an attempt to shape the conservation portion of that land use allocation, providing connectivity among conservation lands and even, perhaps, circumscribing human settlement in a way that will constrain its future expansion as part of a growth management strategy. Fortunately Florida still has the opportunity to protect large intact landscapes within a connected reserve network. We are making great progress towards meeting this goal for the benefits of future generations of Floridians and Florida's biological diversity and natural heritage. However, full realisation of a Florida Ecological Network will depend on the will of the people and politicians and the expertise of scientists and planners through myriads of policy decisions at the local, state, and federal levels over the next several decades.

14

The ecological network development in the Yungas, Argentina: planning, economic and social aspects

14.1 Introduction

The degradation of the subtropical forest in the Argentine north-west can be considered as an example of the effects of neo-liberal policies on nature conservation. These schemes refrain from any attempt at land use planning. What is more, these policies determine profound alterations in this mountain rainforest, which will strongly affect regional sustainability.

The Yungas ecoregion and the Paranaense forest are the hotspots of bio-diversity in Argentina; the Yungas is the more extensive one. There are about 5 000 000 hectares of rainforest in north-west Argentina. As an example, the Yungas host 60% of the total number of bird species of the country as well as some particular mammals such as the jaguar (*Panthera onca*) and the Andean taruca (*Hippocamelus antisensis*). The latter is a mountain deer nominated as 'national natural heritage'. At the same time, this subtropical rainforest is the water producer and the natural water basin for a vast irrigation region. Moreover, the Yungas is a huge carbon sink consisting of extensive forests grading into Andean grasslands at the highest altitudes (Figure 14.1). However, the condition of the Yungas is changing quickly, as indicated by the fact that an estimated 1 250 000 ha have been converted from nature into agriculture between 1975 and 1988 (Reboratti 1989). In this chapter we present an assessment of the changes in this conversion process, and discuss the effects of forest fragmentation on the connectivity of forest patches.

Parques Nacionales de Argentina is trying to counteract this development and seeks to preserve the continuity of the natural processes and the forest's spatial connectivity. We present the land transformations and their impact on society and the environment, and analyse the coherence of the various expectations about land uses. Our analysis will consider a trade-off between

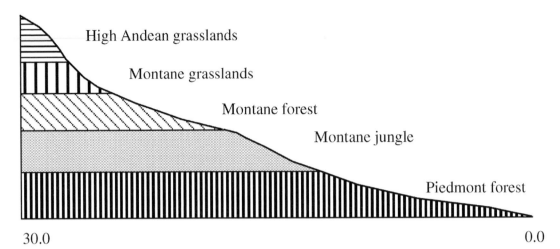

FIGURE 14.1.
A typical Yungas transect and its vegetation zones.

conservation of natural resources and expansion of agriculture. Then, we present a review of the socioeconomic context. Finally, we sketch a prospective view of the Yungas ecological network and regional sustainability. It includes key features of the local community organisation, its potentials and its limitations.

14.2   Land cover changes in the Yungas region

14.2.1 Land cover changes

In the Yungas ecoregion there are two National Parks, Baritu NP (73 000 ha) and Calilegua NP (76 000 ha). The balance between natural processes and productive activities (farming and forestry) is changing. We made a spatial–temporal analysis taking stock of agriculture involving 952 224 ha in five departments (counties): western Oran, Iruya and Valle Grande, western Santa Victoria and the north and central zones of Ledesma (Figure 14.2). To assess the expansion of the agricultural 'borderline' during the past fourteen years, a visual interpretation of Landsat TM images in 1986, 1997 and 2000 has been performed.

In the quantitative analysis only clearing areas have been included that cover more than 5 ha. We exclude those areas dedicated to extensive ranching and selective wood extraction. The classification and interpretation (scale 1 : 100 000) of the satellite images was carried out by signature analysis.

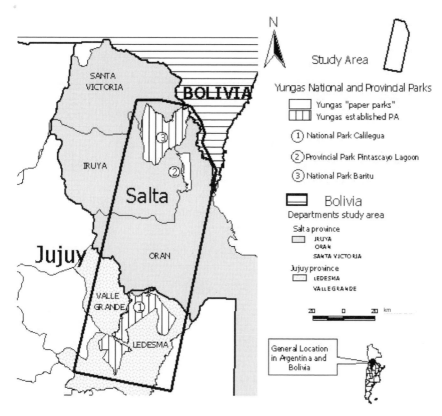

FIGURE 14.2.
Departments in the Yungas region included in the ecological network project. The
National Parks are indicated by the numbers 1 (Baritu NP) and 2 (Calilegua NP).

Our approach has four main components:

1. GIS-based (Erdas Image and ArcView) assessment of textures, shapes
   and colours were used to differentiate five key landscape elements
   depicted in Table 14.1.
2. Analysis of the association between the GPS ground spots and their
   description (in co-operation with INTA Yuto station).
3. Analysis of regional soil maps (Nadir and Chafatinos 1990) and
   cadastral maps.
4. Expert interviews in the Yungas region for additional background.

Half of the investigated area involves soils suitable for agriculture and livestock activities (Table 14.2). The proportion of farming areas (> 5 ha) to 2000
is nearly 12.14%. In relation to 1986–97, the clearing rate doubled in three

Table 14.1. *Land cover categories in the Yungas project*

Class	Description
1	Native forest in good conservation status
2	Native forest in process of increasing degradation
3	Recovering degraded areas
4	Old riverbeds in first stages of conversion to agriculture
5	Agricultural clearings

Table 14.2. *Total agriculture and land used for farming in the study region*

Category	Value
Total area	952 224 ha
Potential agricultural area (PAS)	
(soil classes II, III, V)	481 724 ha
Farming soils in the region	50.58%

Source: The assessment of the potential agricultural area is based on Klingebiel and Montgomery (1961).

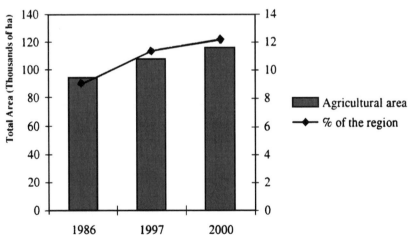

FIGURE 14.3.
Development of agricultural land use in the study area.

years from 1997 (Figure 14.3). These rates clearly show the regional trends of a progression of the 'agricultural' borderline westward. It means the agriculture is going beyond the primary core areas on the banks of the San Francisco River.

14.2.2 Observed spatial patterns in the different departments

In the Valle Grande and Santa Victoria no inference can be drawn from the Satellite images. There are steep slope areas where extensive ranching is the dominant and traditional land use. All farmers use the native forest as the breeding area. These zones have no agricultural potential and the farmers' tracts are not extensive in regional terms. Clearings for subsistence farming are very small, often less than 2 ha.

In Iruya, soils suitable for cropping can be found in zones where the native forest is the dominant landscape matrix (Table 14.1, class 1). Relatively isolated, small glades, mostly less than 5 ha, can be found in a recurring spatial pattern. The glades are located to one or both sides of dissection lines such as roads or rivers. The total area is small on a regional scale but it would be a failure to neglect its role in management of natural resources and development of conservation policies. These small gaps call for special attention. They are potentially a big risk because they are dormant farming nuclei. During the past three years these nuclei have increased in area by 36% (including only gaps > 5 ha).

The incorporation of new agricultural lands results from two trends: the expansion of current gaps and the opening of new ones. It is crucial to consider the location of new gaps between the two national parks. Moreover, if we consider that only 1.43% of the farming soils in that intermediate area was already deforested by 2000, then it is clear that the creation of new clearings will cause partial isolation of the parks. They will be surrounded by agriculture or fruit plantations. As an upshot, there will be impoverishment of the two reserves. Furthermore, the spatial design derived from anthropogenic changes shows the high impact of new infrastructure. New access roads are being opened to areas that have remained unspoiled until now. Therefore, the plans for new developments play a decisive role in the region.

In Oran and Ledesma, present patterns of land use are very different from previous ones. Here, agricultural soils on land with very gentle slopes dominate the landscape. Low risk of erosion and large tracts allocated to intensive farming are key features of this zone. The highly human-modified landscape presents three main units (Table 14.1). In this region land can be classified as (1) land cover class 1 (healthy native forest), (2) land cover class 5 (farming systems, including small forest patches), and (3) land cover class 2 (forest matrix including scattered farming gaps of 5–45 ha).

The changes in the farming area were evident in this zone (Figures 14.4, 14.5). There was a notable increase in agricultural clearings created by deforestation. This process is accompanied by a limited growth of urban areas. Generally, the new croplands were added to the existing ones. The remnant

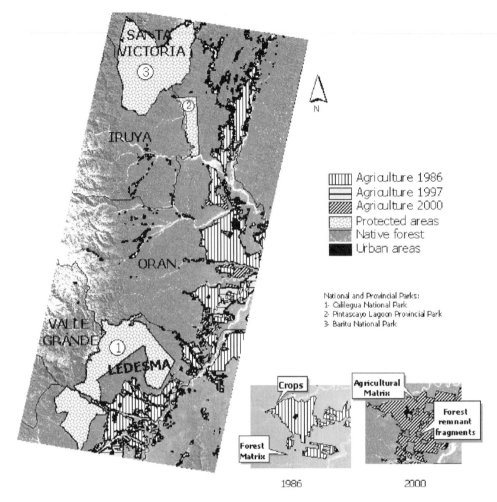

FIGURE 14.4.
Land use change in the period 1986–2000. 1, Baritu National Park; 2, Calilegua National Park. The patterns east of the national parks indicate land transferred into agricultural land use, 1986, 1997 and 2000, respectively. The insert (right) shows the inversion of the matrix during the period considered.

forest patches left over in the merging of the great clearings degraded. Here areas originally dominated by the forest matrix have been cleared and the pattern has been inverted. The matrix is now agricultural with some remnant forest patches. This process determines the decline of effective natural corridors for wildlife.

Ledesma was an historic sugarcane area. This crop corresponds to one of the oldest farming traditions in the Jujuy province. It is the most relevant

FIGURE 14.5
Evolution of the conversion
process for Iruya, Ledesma
and Oran departments.
(A) Development of deforested
areas in hectares for, respectively,
1986–97 and 1997–2000 for the
three regions.
(B) Development in the
percentage of total area suitable
for agriculture.

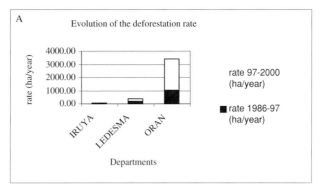

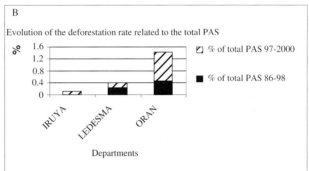

production in terms of area and economic importance. The ecological conditions in this area are most suitable for sugarcane crops, which reach yields well
above the national average.

The new areas added during the study period (1986–2000) at a rate of 194 ha
per year correspond to citrus plantations. There is a worrying expansion trend
from the valley to steeper slopes. The new plantations surround the Calilegua
NP from the east, going up from the San Francisco Valley. These plantations
are mostly lemon and orange orchards. These are high-quality products aimed
at international markets. Other tropical crops are also increasing, such as av-
ocado and mango. The inclusion of these new crops marks a change in traditional agricultural trends: farmers are setting their sights on new markets as a
result of the crisis in the international sugar market. In the period 1986–2000
agricultural area increased by 7.6%. In 1986–97, the clearing rate was 201 ha
per year compared with a rate of 169 ha per year between 1997 and 2000. By the
year 2000 more than 50% of the Potentially Agricultural Soils (PAS) had already
been allocated to farming.

In this department, there are two contrasting farming systems. These are
high-tech large corporations and small farms (minifundios) surrounding the

southern and eastern limits of Calilegua NP. The rate of increase is not faster because there are some structural limitations related to the traditional management systems. This is not only true for the small farms but also for the large corporations. Both share poor management of soil, leading to soil degradation, and an uncaring use of the forest resources.

The Oran department represents a great variety of image patterns and farming systems including sugarcane, grain crops, vegetables, and tropical crop plantations. There are zones with farmed tracts of 600 m × 600 m, others with sides between 600 and 250 m and others less than 250 m × 250 m (mostly around urban areas). In the north-eastern part of the department the pattern is similar to that in Iruya. Vegetables, chiefly tomato (J. Wallberg, pers. comm.), account for a large share of agricultural land. While citrus plantations are increasing, banana plantations are diminishing. The dynamics of the productive landscape have profound social effects that will be discussed later.

The principal zone crops occupied the additional lands (this addition was at a rate of 1334 ha per year). In 2000, farms accounted for 32% of the PAS. There is a frightening increase in the clearing rate between the first (1986–97: 1061 ha per year) and the second period (1997–2000: 2333 ha per year). During these fourteen years farming area increased by approximately 32%. Projections of current clearing rates indicate that it will take less than 68 years to clear all the PAS. This process will mostly affect the piedmont forest and will also have consequences for the transition forest. Substantial sections of these forest types are lost each year, as they are the preferred clearing zones because of their gentle slopes and their potential for agriculture. The associated fauna are losing their habitats, with obvious detrimental effects on the parks. What is worst, Oran is characterised by logging activities. Oran city sawmills mainly process native woods. State subsidies, however, are encouraging eucalyptus reforestations, which could help to buffer the native forest.

This analysis highlights the pressing need for fast and effective management action. This is essential for the sustainable development and conservation of natural resources in the region. There remains about 76% of PAS by traditional methods. The limitations on expanding the agricultural areas are generally related to management systems and technological levels. In other departments of the region, new cropping systems in terraces are being developed.

In addition to the impact of clearance for agriculture, we need to consider other activities. Important extracting activities include the oil and natural gas industries, involving searching, extraction, processing and transportation. In addition, we must consider extensive ranching, enlargements of settlements

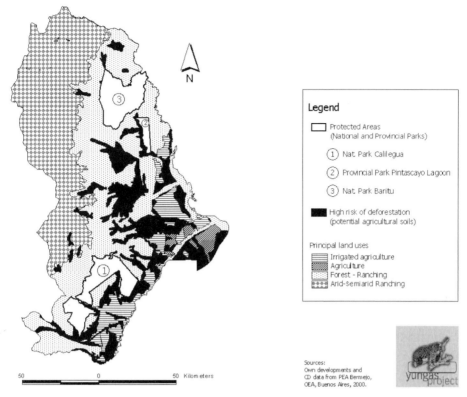

Legend

☐ Protected Areas
(National and Provincial Parks)

① Nat. Park Calilegua

② Provincial Park Pintascayo Lagoon

③ Nat. Park Baritu

■ High risk of deforestation
(potential agricultural soils)

Principal land uses
Irrigated agriculture
Agriculture
Forest - Ranching
Arid-Semiarid Ranching

Sources:
Own developments and
CD data from PEA Bermejo,
OEA, Buenos Aires, 2000.

FIGURE 14.6.
Current land use in the Yungas region, and areas at potential risk of deforestation.

and urbanisation as well as illegal hunting. All these are severely disturbing the continuity of natural processes in the Yungas region. The risk of further deforestation is a latent threat to the regional connectivity of native forest patches (Figure 14.6). These new gaps will in general increase further the isolation of the parks. How far could agriculture be extended up the hills? This is one of the aspects dealt with in the following connectivity analysis. This outcome is a key for planning in the matrix between the parks.

14.3   Socio-economic aspects

14.3.1 The socio-economic structure

The population analysis must account for the principal aspects of current socio-economic issues. The socio-economic situation is a relevant aspect of the conditions in Salta and Jujuy. Both in history and at present these provinces are one of the most undeveloped zones of the country in both productive and socio-economic aspects. The dominant position of farming makes the economic output highly dependent on dramatic oscillations in the rural sector. The fact that production of commodities does not involve much aggregate value adds to the economic instability of local industries. The current national recession, coupled with neo-liberal policies at both national and provincial levels, has deep social effects. Some of these effects are very low salaries and high unemployment rates. This leads in turn to social unrest, and to a government trend to step up the use of violent means for its control. A difficult financial situation for the government is shown by the tight balance between the value of total income and expenditure.

The current global trend is to funnel a larger proportion of gross national income into the richest sections of the community. In regional terms there is also a tendency to a larger concentration and greater dependency in the metropolitan area around Buenos Aires. Large corporations are increasingly important actors in the production of commodities. They have obtained new land and often lead in the development of new crops. These structural aspects are common to all the departments analysed. However, it is possible to differentiate them by the investment of private capital. Therefore, we define two groups:

Group 1:   Those departments that favour the entry, establishment and expansion of past and present agrarian–industrial capital. The lowland departments of Ledesma and Oran compose this group.

Group 2:   Those departments where subsistence ranching and agriculture are dominant features of current farming structure; they are the mountain-dominated departments of Valle Grande, Iruya and Santa Victoria.

14.3.2 Population distribution

The analysis of the inter-census variation of departmental and provincial populations indicates, with few exceptions, a high rate of yearly growth. The rate of population increase in the past two decades has increased markedly (Figure 14.7).

This growth is strongly connected with a higher crude birth rate, related in turn to the low degree of socio-economic development. This is associated with two causes. The low-income population has limited access to family health benefits (e.g. contraception methods). There is also a need for larger families as a source of farming hands. Agricultural tasks occupy all the family on the small farms.

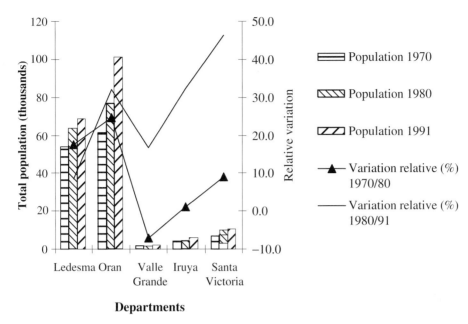

FIGURE 14.7.
Size and growth of total population by department in 1970, 1980 and 1991.

Let us consider the principal difference. Ledesma and Oran show a higher growth rate than the departments that form Group 2. However, for the last period the growth of Ledesma is slowing down. Valle Grande had a negative growth rate in the first period considered, because of migration to more dynamic areas. These were the agricultural industries from Group 1. Migration follows a work chain from the agrarian industries (mostly the sugarcane factory) to the services sector.

In addition to the importance of the rural sector in the region, the statistics show a dominant rise in the urban population. The weight of this sector has been increasing in the last decades. In comparison to the Pampas region, the wealthiest in the country, the urbanisation process in the north-west is much slower. Urbanisation trends are, none the less, clearly substantiated by the data in Figure 14.8. The intermediate size cities (30 000 – 80 000 inhabitants) are attracting people from the countryside. This means that the growth of urban populations is explained, to a great extent, by a drop in rural population. Both aspects can be inferred from Figure 14.8.

These variations have different causes. We consider that this is partly related to modernisation of agriculture and the concentration of land ownership. These changes are strongly associated with the increasing role of large agribusiness corporations, which have mechanised most rural activities in

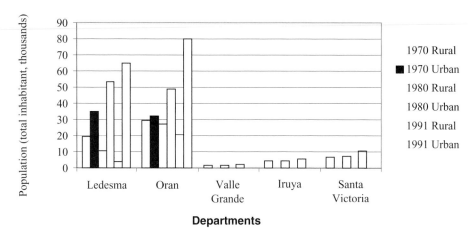

FIGURE 14.8.
Size of the rural and urban population by department, 1970, 1980 and 1991.
Source: based on INDEC (2000a, b).

recent years. Such mechanisation causes unemployment, especially among the temporary workers. In the sugarcane production process, mechanisation results in the displacement of 70 workers by a sugar harvester. These populations have migrated to the nearby towns. This case is characteristic of Ledesma, where the unemployment rate is about 7.5%, the highest in the studied departments. Further, the population in this department shows an ageing trend, in comparison with other departments (INDEC 2000a). It could be related to migration of younger people to the provincial capital or to other urban centres in search of better employment opportunities.

Most of the migrant people come from other provinces and neighbouring countries, especially from Bolivia. Some of them have found jobs in the rural sector. The people displaced by the mechanisation process have tried to find opportunities in the urban centres. Because of their dynamics, the principal destinations are Ledesma and Oran. The national economic crisis, which hits harder in impoverished regional sections of the north-west, motivated some of the subsistence farmers to migrate to the urban centres. Some of them have tried early vegetable production as an alternative. They have settled in the vicinity of the Group 1 towns to develop this activity, targeting the nearby consumption and commercialisation markets. Other former farmers have joined the public sector as civil servants. This is a characteristic of the developing and poorer provinces (Cao *et al.* 1997). We note this tendency in Group 2 departments (Figure 14.9).

Employment structure

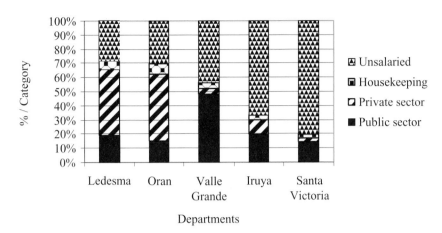

FIGURE 14.9.
Distribution of the employed population by working category by department.
Source: based on INDEC (2000a, b).

The data for urban population are presented only for the departments of Ledesma and Oran. The other three cases have no urban population in terms of the Argentine census methodology. The national census considers urban population as those living in places with more than 2000 inhabitants. The villages in the Group 2 departments have smaller populations.

This situation is clear in Valle Grande, where it is possible to see a relevant escalation of the so-called rural grouped population (small towns with fewer than 2000 inhabitants). In the period 1980–91 the increase was 593% (Manzanal and Arrieta 2000). Most of these people developed transhumance ranching and subsistence agriculture, and then abandoned the farm to take public sector jobs. These jobs are deeply related to the political parties' favourism (Levy *et al.* 1999). In this group of departments the private sector is important because of the corporations and the service activities.

The opening of the national market allowed the import of rural products. A number of products of the local farmers cannot compete with the prices of foreign products. In the area of Oran, the banana plantations diminished from 7000 ha to 4630 ha in a 9 year period (1987–96). The products imported from Ecuador wiped out the local plantations and contributed to high rural unemployment. This was absorbed by the city of Oran, where precarious settlements were developed hosting about 32 000 people. This process lasted about four years. It caused a severe social problem because of unemployment and marginality, and overloaded the city services and the social assistance network (Arrieta and Pastor 2000).

14.3.3 Quality of life

The recent economic cycles have had a negative effect on provincial economic growth because of the growth of unemployment as a product of both the national economic crisis and dominant policies that hinder economic activities. It deepened the social differences within each province. Nevertheless, production of commodities increased through a marked expansion process, led by agricultural business, boosting the incorporation of new lands and new crops. But the problem is that this increase in agricultural areas has been focused in sectors of highly concentrated capital. These sectors do not invest their earnings in the region. This means that the process has not promoted the development of most of the provincial population, nor did it contribute to the enlargement of the labour market and improvement of family incomes.

The enlargement of the public sector (as government surrogate) constitutes a strategy to diminish unemployment. However, because of its underlying structural causes a significant proportion should be considered as 'shaded unemployed' or sub-employed (Manzanal and Arrieta 2000). We consider that this type of employment does not solve the arduous regional situations of unemployment, uncertainty, and associated social unrest. These public jobs do not constitute a genuine source of permanent profit with positive effects in the region. This is particularly evident in the Group 2 departments (Table 14.3).

Table 14.3. *Population characteristics concerning housing, medical and social situation, and illiteracy.*

Analysed from data provided by Manzanal and Arrieta (2000).

Administrative political division	Population without social assistance or medical plan or insurance (%)	Houses that are substandard (%)	Illiterate population (%)
Jujuy total province	43.5	45.0	6.7
Salta total province	47.8	45.4	6.7
Group 1			
Ledesma	39.6	36.6	7.0
Oran	55.6	63.9	10.0
Group 2			
Valle Grande	72.2	92.8	19.8
Iruya	70.5	95.6	20.6
Santa Victoria	79.8	95.6	29.2

14.4   Our expectations and the need to co-ordinate actions

There are some key issues to be solved for a comprehensive development of the Yungas ecological network. The technical aspects related to the connectivity between the reserves (to make fauna movements easier) are not fully elucidated. We are at present researching the suitability of different methods (graphs, cellular automata, etc.) to analyse regional connectivity. The purpose is to detect the most 'strategic' forest patches and corridors (in the matrix, among the existing protected areas) in order to focus conservation efforts there. Besides, the legal framework related to nature conservation in Argentina does not yet recognise the need for different management categories related to the elements of the ecological networks (natural corridors, breeding zones, etc.). It considers only the different classes of protected areas (National Park, nature reserve, etc.). Therefore, an updating process would be very useful to add legal support to the development of the required elements.

The other issue nowadays in the international conservation arena is the need to create closer and deeper relationships with the stakeholders, i.e. to form alliances with the local population (Mwamfupe 1998; Cowling 1999; Ntiamoa-Baidu et al. 2000; Margoluis et al. 2000). We decided to take part in this discussion. Since January 2000 we have broken with the tradition and history of the Argentina National Parks Administration (APN, Administración de Parques Nacionales). The starting point was to work together with the local communities in the whole country and to stress the co-operative relationships with other state agencies. This policy has been put into action with different local groups. In particular, the Parques Nacionales carried out 'new conservation policy guidelines' in the Yungas region. The Superintendents of the Calilegua and Baritu NPs supported this strategy. The first action was an important milestone, a workshop about 'Indigenous Protected Lands'.

The approach brought new visions on nature conservation, local communities' engagement and indigenous knowledge in support of the Parks federal administration. At the same time, this new approach produced a kind of conceptual earthquake inside the Agency. The Argentine parks were created long ago in 1932 and APN has kept its traditions. Only in the period of presidency of Jorge Morello (1983–89) was a real effort made to promote the participation of provinces and local communities. At that time APN tried to develop effective partnerships. Considering these antecedents, both the people and the APN staff used to think and act as closed compartments. It was a cultural and social apartheid that brought conflicts and hard political discussions. Today, this is changing slowly during a process that requires time and continuity.

14.4.1 The case of Estate Santiago and Estate San Andres (Tinkunaku community)

We realised that the conservation of the intermediate space between Baritu and Calilegua would not be possible without the support of local communities. Fortunately, they share our conviction about the need of conservation of these lands. This is related to their singular valuation of the forest as a whole. There is a deep difference with the western world. The Kolla people have a strong attachment to the forest. Moreover, they consider themselves to belong to the land and not vice versa. They firmly believe that without the forest the complete region, and their lives, will vanish. There is a kind of empiric and ancestral sustainable vision. The Kolla people from the Argentine Yungas have two principal communities: Estate Santiago and Estate San Andres (Tinkunaku community). They have one council in each Ayllu (village). All the local councils constitute the Assembly. There are four different local councils in each principal community. These communities lack a vertical organisation; everything is discussed first in the regional Assembly.

These two communities have been fighting legally for their rights of land ownership since 1940. Recently, Estate Santiago has received the official ownership titles from the federal and provincial government. The Tinkunaku community, however, is already in a new legal battle with one of the sugarcane corporations. They have received only about 19 000 ha out of the 125 000 that they are claiming. This corporation claims also ownership of the land.

14.4.2 Production, conservation of the ecosystems and land ownership

The Kollas (Figure 14.10) developed transhumance as a special way to take advantage of the ecosystem diversity (Dominguez and Mariotti 2001). This production system implies not only the seasonal movement of the cattle herds but also the carrying of families, tools and portable goods. Their circuit is from the lowlands in winter to the highlands in summer. The Kolla people define the transhumance as 'their method of life' (C. Zarate, pers. comm., 2000). This way of life requires the whole environmental gradient. From the Kollas' point of view, their existence and functioning as a community is only possible with the whole set of ecological zones that is present in the Yungas from piedmont forest and mountain forest and valleys to the Puna, the high Andean grasslands (see also Figure 14.1). The social structure of the Ayllus (groups of communities) has its origins in the Inca communities' traditions (Untoja, cited by Dominguez *et al.*, undated). It has a deep cultural sense in relation to the central religious meaning of Mother Earth – Pachamama – the mother of all of us. Land and people are seen as a single entity. The Kolla are in essence seasonal itinerant cattle

FIGURE 14.10.
The Kolla people and their Yungas.

shepherds. Their economy is based on small tracts for subsistence crops and extensive ranching. However, they have many contact points with western culture because of the provincial education system. They are managing computers and they are developing an electricity network based on solar panels.

14.4.3 The forest and grassland conservation

We know already that there could be some minor readjustments in the Kolla system. Sometimes, they use fire as a tool to enlarge the summer foraging area. There is no major problem with the fire itself, but some forest types have been lost owing to these practices. These losses are, however, negligible in comparison with those caused by the deforestation process in the valleys. None the less, it is important to also evaluate these effects while keeping mind that the forest and the sense of careful management are pivotal to the Kolla culture. They have even organised their own reforestation programme in an old degraded forest, using the motto of 'four new trees for one cut tree'.

14.5   The future of the Yungas network

The nature conservation task in the Yungas is not an easy one, because it deals with the whole setting of nature conservation and socio-economic conditions. The Yungas are a high biodiversity hotspot, a water pump, a rich irrigated cropland, a source of hydrocarbons (gas, oil) and a very valuable cultural space. It is a theatre of dissimilar interests. Considering some of these facets, the work with the local communities is essential. Nevertheless, it is not a panacea. The designing of an ecological network leaving adequate prospects for agriculture and to counteract unwanted consequences demands tailor-made regional and local plans (Jongman *et al.* 1995; Davis *et al.* 1999). Hence, the team of Parques Nacionales de Argentina (APN) is trying to develop closer relationships with the different interest groups. There are huge corporations, small farmers, different state agencies, and provincial governments. All these sectors have singular perceptions, driving forces and methodologies. Together with the two environmental provincial agencies we defined a first tool to solve this puzzle in the Yungas biosphere reserve project. It is our common arena to promote an open discussion and seek for consensus on nature conservation and social and economic development. We all know that it will be a long-term process towards sustainable land use in both economic and ecological senses.

We need to face all challenges; this is a key point of the new APN policy guidelines. This new approach has already obtained some results. Salta has its first law relating to the provincial protected areas system. There are park ranger candidates from the NW provinces at the APN Rangers School (from Salta, Jujuy and also from the Tinkunaku community). In addition, two new reserves have been created in the past two years. Both are very close to the Baritu NP. One is the first Salta provincial park: Pintascayo Lagoon (13000 ha) located in the south-eastern border of Baritu NP. 'El Nogalar', an 8000 ha APN natural reserve, was recently created NW of Baritu, very close to Toldos town where Baritu NP Rangers have their base station. Moreover, a recent national law declared the jaguar (the Yungas flagship species) as 'national natural heritage' (September 2001). It constitutes a special call for the conservation of its habitat, the rainforest.

The Yungas ecological network needs this kind of effort to become a reality. Now, APN is working very closely with the provinces and all the other stakeholders to solve the 'balancing problem' (van Lier 1995; De Lima Pufal 2000). A consensus vision about the different regional perspectives must be the way to get participating future scenarios. It is the way in which we are building the future.

15

The river corridor of the Guadiamar

15.1 Introduction

The progressive impact of humans on the European continent for the past several centuries has caused the degradation of a large part of the natural systems, limiting the natural or slightly altered areas to small units, generally located in zones with difficult access or small economic value. This degradation of natural areas has caused an isolated evolution of those ecosystems that are better conserved and where wild species of fauna and flora develop. In many cases the geographical disconnection so produced has provoked the extinction of species or the diminishment of their populations to the point of endangerment. Another effect of this geographical isolation is seen in the limited dispersion of the animal species and the endogamy that is produced in some cases.

The Doñana area is an example of this isolation phenomenon, and if a large part of its territory from the past still exists, it is because of the small or null economic value that this land has had. On the other hand, its ecological characteristics have enhanced the conservation of its ecosystems. Once the Doñana territory has been considered protected, it is important to consider the problems of its isolation and disconnection from other natural territories that long ago were interconnected. The Guadiamar river constitutes an element of considerable value as a link between two great natural systems: the Doñana and the Sierra Morena mountains.

We can consider the Guadiamar basin as an area where culture and nature are integrated but which over time has been losing value in both senses, and which surely has been greatly devaluated since an important mining accident happened in April 1998. The recuperation of this landscape, identified with Western Andalusia, may mean the creation of a model ecological corridor for flora and fauna. In addition, it means a new dimension in management approaches for protected natural spaces, where restoration is supposed to be a priority issue for the recuperation of lost and forgotten landscapes.

15.2   The Guadiamar River and its catchment

The Guadiamar River is one of the most important tributaries of the Guadalquivir river and has as main tributaries the Agrio, Ardanchón, Alcarayón and Cigüeña rivers. The total catchment area of the Guadiamar measures 1879 km². The length is 83 km and the source is at an elevation of 320 m. The Aznalcóllar mine complex and its tailings pond are placed on the right riverbank of the Agrio river, 2 km upstream of the confluence with the Guadiamar river (Figure 15.1).

The Guadiamar basin is part of the Low Guadalquivir System, which plays a relevant role in the ecological integrity of the larger Doñana system. It is located on the north-western edge of the Guadalquivir river basin and is the core for the Green Corridor. Three essential sections must be distinguished, which extend within physically distinct territories.

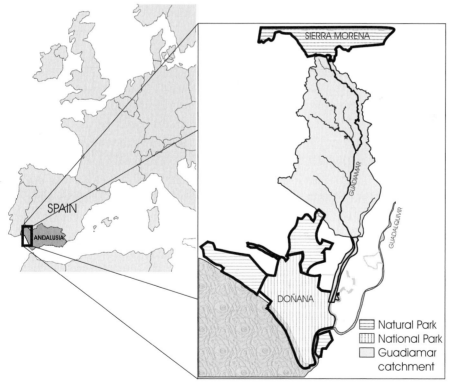

FIGURE 15.1.
Location of the Guadiamar river and its catchment. The location of the Aznalcóllar mine complex is shown by the symbol.

1. The first or north section is an extensive territory that unites the Guadiamar river with the limit of the Natural Park of Aracena and the Picos de Aroche through natural areas of great ecological value with pastures and typical Mediterranean forests. It is the drainage area of the river system. Within this mountainous terrain, the river course is clearly delineated. Erosion dominates in this area and sweeps away the sediments.

2. Having passed the spurs of the Sierra Morena, it flows through an extensive valley in which the river forelands get flooded during the rainy years. Three terrace levels have been recognised along the Guadiamar valley, each one being made up of a lower layer of gravels and an upper one of sand and silt (Manzano *et al.* 2000). The actual river channel is an erosive form excavated in the lowest terrace. This central section, where the mine spill affected the area considerably, reaches from the Aznalcóllar mines to the start of the third zone, called Entremuros.

3. The lower stretch constitutes the third section or south section. In former times its water flooded the marshes of Doñana but since the 1950s the river has been confined to its bed by two artificial earth dikes down to its mouth into the Guadalquivir river (Grimalt *et al.* 1999). The area in which the river is channelled is known as 'Entremuros' (between dikes) and represents the altered river system.

From a climatological aspect, the Guadiamar watershed can be classified as a subhumid Mediterranean climate region of heavy irregularity. The Guadiamar River has three gauging stations for discharge measurements and within its basin there are 15 rainfall stations. Thus datasets could be used to model the rainfall on a regional basis (Velasco *et al.* 1991). For the southern gauging station the mean flow rate is $6.6\,\mathrm{m^3\,s^{-1}}$ ($208\,\mathrm{hm^3}$ yearly) with a standard deviation of $5.3\,\mathrm{m^3\,s^{-1}}$ (Gallart *et al.* 1999). This standard deviation means a high interannual irregularity: some years have more than $300\,\mathrm{hm^3}$ discharge whereas other years have less than $50\,\mathrm{hm^3}$. The high discharge period is from January to March whereas the low discharge period is from June to October.

For several hundreds of years the Guadiamar watershed in general and its river network in particular have suffered a serious process of degradation (Arambarri *et al.* 1984). This has caused a serious ecological deterioration of the river bed and the riparian wetlands, some stretches of which are now in a deplorable state. Currently, the most important causes of this situation are not only mining activities but also the waste water from the larger villages located in the valley and waste water from some agribusiness industries in this area. Moreover, there is almost complete deforestation of the riverbanks for agricultural purposes (Prat *et al.* 1999). The terrestrial ecosystems of the watershed, characterised by various Mediterranean trees and bushes (holm-oak woods, wild olive trees, pine groves...), have been reduced essentially to small unconnected fragments among a wide agricultural matrix of herbaceous crops, which are grown by either dry or irrigation farming.

15.3   The Aznalcóllar mine accident and the emergency measures plan

In the spring of 1998, a very serious mining accident took place at the Aznalcóllar mine. One wall of the pyrite-mine tailings reservoir, which contained up to 25 hm^3 of pyrite-rich mud and acid water, collapsed and caused a spillage of 6 hm^3. Between the broken dam and the Entremuros area the spillage affected a strip 62 km long of the Agrio and mostly of the Guadiamar rivers, covering an area of 4634 ha of riverbanks and adjacent farmlands (Figure 15.2). Because two transverse earth dikes were built at Entremuros a few hours after the accident, the floodplains were inundated and polluted over a length of only 13 km.

Owing to their fine particle size, after the accident the tailings were easily transported in suspension in the flood wave. Most of the mud was deposited in the first few kilometres downstream of the tailing dam. The maximum thickness of the deposit was 4 m in the first 5 km downstream close to the reservoir, whereas in the farthest 5 km it was less than 5 cm thick, a few millimetres near Entremuros (Querol *et al.* 1998). Its content was mainly residues from the pyrite extraction processes, with a fine-grained structure consisting mostly of sulphides, pyrite being the dominant mineral (75–80%), and some silicates. This means that the concentration of heavy metals was high.

Acid water (pH 2–3), with various dissolved metals, toxic to living organisms and a potential pollution source for groundwater in the flooded alluvial aquifers, and some of the finest colloid-size sulphide particles, arrived at the Entremuros area, where they were confined between dams.

The administration's major concern was to adopt immediately efficient measures to minimise the effects of the accident and to avoid the spread of contamination. A special co-ordination centre for the follow up of the emergency was set up, and carried out the following actions: defining the characteristics of the spillage, monitoring of flow levels and flooded areas, sanitary control of water and food, determination of affected agricultural areas, and immobilisation of agricultural products.

One of the key decisions was to remove the polluted sludge before the rainy season in order to avoid serious problems. Thus, a few weeks after the spill the sludge began to be removed from the affected area and carried to the already abandoned Aznalcóllar open pit mine, once reports by the National Geomining Institute had confirmed pit impermeability. This huge task involved more than 500 people, 450 trucks and 250 different machines per day; the total amount of material removed was over 7 hm^3. The clean-up finished in November 1998.

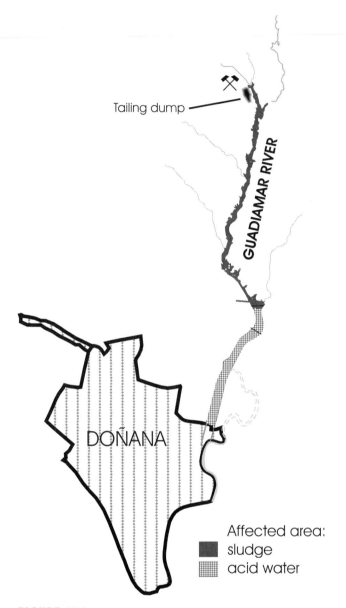

Tailing dump

GUADIAMAR RIVER

DOÑANA

Affected area:
sludge
acid water

FIGURE 15.2.
Location of the Guadiamar lower course, downstream of the tailing dump, showing the area that was covered by sludge and the area that was only covered by acid water.

Another important action, in August 1998, was the chemical treatment *in situ* of retained water at Entremuros (up to 4 hm³ at that moment) before its controlled release into the Guadalquivir river. After that, the superficial soil layer and vegetation at Entremuros had to be removed owing to the contamination incurred as a consequence of polluting water retaining mainly zinc precipitates.

At the same time monitoring and characterisation works also started, carried out by a number of different research bodies. The work included:

- mapping of the sludge spatial distribution and thickness
- analysis of the sludge mineralogy and geochemistry
- column tests to determine the transport parameters of heavy metals in soil
- groundwater sampling to characterise regional hydrochemistry and to study how and to what extent the aquifers were affected by the acid spill
- monitoring of the impact on the wild fauna and flora

The accident did not cause personal injury, but the socio-economic effects were also important. More than nine townships in Seville Province were affected because they lost agricultural crops and mining activity had to be stopped. Moreover, indirect effects, such as potential health risks, the impact on the local image and the devaluation of regional agricultural products on the international market, played an important role.

All these aspects are important to determine the predicted impact of the residual contamination that could remain after the sludge cleanup. This would affect agricultural production in the near future and was the origin of one of the key decisions that the regional government had to take: farming was forbidden on the affected land and its ownership was acquired by the regional administration in order to realise its ecological restoration. It is worth highlighting that the change in the use of the farm lands affected by the mining spill, by incorporating them as part of the Green Corridor, safeguarded the certified quality trade name of the farm products from the non-affected areas. This reflects the need for security and guarantees with respect to the national and international markets.

Ultimately, the objective pursued was not only to eradicate the unwanted bad reputation of the Guadiamar River but also to reverse this situation so that people would no longer associate this name with a serious mining accident. The objective was rather to show that, despite suffering a serious ecological disaster, this area was capable of implementing a sound, realistic, advanced and innovative project of ecological restoration and territorial reactivation, which could even become a relevant model for other places in Spain and in the world.

15.4   The strategy of the Guadiamar Green Corridor

The project of the Guadiamar Green Corridor was orginated and is being implemented by the Andalusian Regional Government. It is based on the conviction that it is necessary to integrate both the restoration and conservation of the ecosystems of the Guadiamar watershed (including its biological diversity) and the sustainable development of the associated human systems.

A central objective in this approach is the creation of a fluvial corridor with, from the beginning, two well-defined aims. It is intended to achieve gradual recuperation of the river's natural conditions in order to restore the ecosystems of this Mediterranean watershed and riparian zone. In this way also the dispersal of species and ecological processes will become feasible, as well as the objective of improving the quality of life of its human inhabitants.

An ecological corridor is a continuous natural area in the form of a band of linear vegetation that differs from its surroundings and that unites two or more natural zones (Saunders and Hobbs 1991). In this sense the function of the Guadiamar as an ecological corridor had almost entirely been lost through human use. Thus, the proposal of the Andalusian Government for the integral regeneration of the area affected by the mining accident is the constitution of an ecological corridor. Besides, the potential creation of the ecological corridor would also satisfy one of the most important claims presented by the different green associations of Andalusia, who regard its implementation as a real means for breaking the growing isolation of the ecosystems of Doñana. The natural area under consideration is about 10 000 ha and will include not only the area affected by the sludge but also the area located to the north of the mining complex and which is an indispensable link for the connection with the Sierra Morena mountains.

The project intends to set up an alternative and sustainable development process, also based on new economic activities and opportunities created by the corridor (Elbersen and Prados 1999). The corridor project deals with the management of the natural and human heritage in order to achieve an harmonious and well-balanced co-existence between the exploitation of the natural resources of the river basin for of the general welfare of its human population and the conservation of its biophysical processes.

However, the individual users as such must not be seen as the only possible beneficiaries of the project; it is also necessary to regard both the general public and private sectors as driving forces for social development. The people in Andalusia will benefit from such a strategy, when implemented, because it will considerably improve the currently highly degraded environment of a watershed. It also will give back the great variety of services once offered by its ecosystems, which had been very much reduced or had simply disappeared. Moreover, the public sector will take advantage of the new conditions stemming from an interconnected territory. The institutions involved (national, regional and local administrations, among others) will be in a better position to co-ordinate and achieve their objectives within a co-operative structure of all stakeholders operating in the watershed.

The project is a multi-disciplinary working process carried out by a working team set up by the Regional Ministry for Environment, consisting of technicians from the different offices, researchers from several universities and from the Higher Council for Scientific Research (CSIC), as well as members from green associations. The Strategy of the Green Corridor was published, including all the above questions, and conceived as a document with sound scientific grounds, being technically feasible and economically viable and with the necessary social and political support.

For the development of a high degree of social support as regards its objectives and basic action lines, a long-lasting process of people's participation is needed. The involvement of all the institutions and society in general justifies a public participation procedure for the development of the Green Corridor Strategy. This process has been designed to ensure that the strategy can be something more than just a plan worked out in an isolated manner by groups of technicians and scientists. This approach is used to develop a process of co-operation that will also take into account the aspirations and priorities of the various institutions and different socio-economic groups that play a particular role in the future of the region.

With the aim of attaining an initial scientific and technical consensus, an International Seminar on Ecological Corridors and Restoration of Rivers and Riparian Zone was held in Seville in June 1999. The Green Corridor Project was presented and discussed at this seminar and later in different public meetings that were organised to achieve, on those occasions, a social and political consensus.

To achieve the main objective of the Green Corridor Strategy a basic Action Plan was designed. It consisted of four main worklines that govern the efforts made to rectify the contamination from the mining spill and restore the degraded aquatic and terrestrial ecosystems of the Guadiamar fluvial system and watershed in order to recover its natural function as an ecological corridor. Each workline is aimed at one or more overall objectives to be attained through a series of programmes that, in turn, should achieve particular goals by means of concrete actions. The four worklines are:

- design of the Guadiamar ecological corridor
- remediation of contamination
- ecological restoration
- integration of human and natural systems

The worklines have a scientific basis in a research programme, PICOVER (Consejería de Medio Ambiente 2000). It consists of a series of applied research projects, many of which employ a multi-disciplinary approach, that attempts to satisfy the high level of knowledge required for the design and implementation of an ecological restoration activity like the one at stake. An effective implementation of the Action Plan would require a periodic follow-up and evaluation of its action programmes. Therefore, a monitoring programme has been designed, from the start, for each workline to explore and assess its success or failure, and to collect the necessary information to modify and/or adapt its particular goals and methodological procedures. Figure 15.3 shows the interrelations among these three elements: Action Plan, Research Programme and Monitoring Programme. Finally, in all phases of the development of the Strategy, whether it be during its elaboration, implementation or follow-up, the information, collected and stored in standardised and geo-referred data bases, is spatially organised, managed, analysed and interpreted through the Environmental Information System of the Regional Ministry of Environment (Figure 15.4).

The four main worklines that form the content of the Green Corridor Strategy have been developed in the first of them, which deals with the theoretical design of the Guadiamar river corridor as the main connecting system between the Sierra Morena and the Doñana littoral systems. They finally consider the development and implementation of the ecological restoration programmes necessary for their achievement. Obviously, the main objective of the strategy will be attained if a considerable area of the riverbed and floodplain contaminated by the mining spill is remedied, and if the socio-economic and cultural demands and requests made by the population of the valley are taken into account. The latter aspects will be indicative for the content of the other two worklines.

15.4.1 Design of the Guadiamar ecological corridor

The purpose of the river corridor design is to develop a landscape with great ecological heterogeneity and a high degree of multi-scale connectivity, through physical, biological or cultural corridors or even without corridors, in order to improve connectedness (see chapter 2) by connecting well-conserved fragments of natural ecosystems. This goal will be achieved through management patterns that will encourage the multiple use of goods and improve services from the aquatic and terrestrial ecosystems of the watershed. In this context, we must not forget that the traditional use of natural resources also plays a very important role in the functionality of many types of ecosystem, inasmuch as historically a sort of co-evolution process between natural and cultural elements has been established. Thus, if we promote both traditional and current systems with diversified uses, we could obtain a landscape characterised by great ecological variability and formed by highly connected ecosystems – whether protected or not – in different phases of their

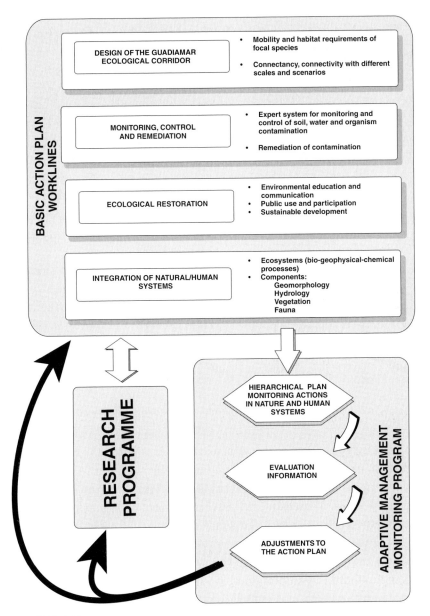

FIGURE 15.3.
Interrelations among the Action Plan, the Research Programme and the Monitoring Programme.

ecological development (Wells and Brandon 1992). We will have well-conserved mature ecosystems, generally protected (Natural and National Sierra Morena and Doñana Parks), but also a whole set of spaces transformed by humans (olive groves, vineyards, pastures, pinewoods, herbaceous crops, etc.) that are also of value for nature.

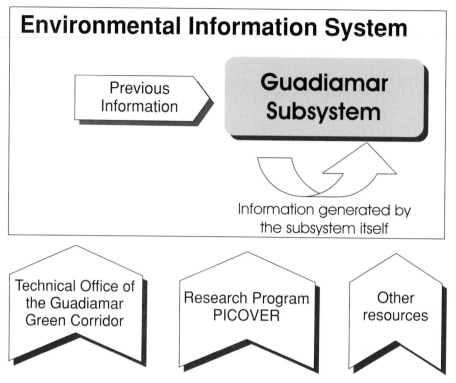

FIGURE 15.4.
Information flow in the Andalusian Environmental Information System.

Although the issue is still controversial, it is currently considered one of the best strategies to increase the connectivity – i.e. the degree of functional connection – of a landscape, through the development of corridors as physical structures. These can link patches of more or less conserved ecosystem fragments within a matrix of degraded or destroyed ecosystems. From such a perspective, ecological corridors are just structures in the landscape that function to improve connectivity, in this way being essential to reduce the effects of an advanced fragmentation process. Fragmentation is one of the causes of the accelerated local and global extinction of plants and animals. Among all types of corridor, river corridors are considered the most important because they embrace both river beds and floodplains. Currently, ecological networks of natural spaces, seen as the quintessence of nature conservation, are being developed as ecological corridors by restoring river systems' functionality. The development of the Guadiamar ecological corridor is the first experiment carried out to give functionality to the Protected Natural Areas Network of Andalusia (RENPA) through different fluvial systems (Figure 15.5).

The design of the Guadiamar river corridor is being developed within the territorial scope of its watershed. In the spatial fabric with a close interrelation

FIGURE 15.5.
Distribution of the natural areas included in the Protected Areas Network of
Andalusia (RENPA). The arrow indicates the Guadiamar Green Corridor project.

between all abiotic, biotic and human components, a dynamic process takes
place linked to the water cycle. The Guadiamar river corridor must be regarded
as a functional unit defined by a longitudinal, lateral, vertical and temporal dimension and formed by four constituent ecosystems: the river bed, the
floodplain, the transitional strip next to the fluvial terraces, and the alluvial
aquifer.

The factors and processes that determine river corridor functionality are
expressed by and operate at different space and time scales. These processes
must be identified and taken into consideration to develop viable conservation, decontamination and restoration programmes. At different observation
and analysis scales the conclusions and results that can be obtained from an action programme can vary a great deal. From such a multi-scale vision, the river
corridor may be seen as a hierarchy of interdependent ecosystems of different
size, where any perturbation (of natural or human origin) will affect the components of the inferior ecosystems. In this way it is possible to plan and manage
it as a whole, as an integrated and unitary entity.

15.4.2 Pollution control measures

For the workline relating to the contamination of soils, waters and organisms, as a consequence of the mining spill, development and implementation of an expert system is needed for monitoring and controlling residual

contamination. A network of sampling stations was put in place along the Guadiamar river bed, floodplain and alluvial aquifer, which, together with the network of the Guadalquivir estuary, will provide this system with basic information.

After the cleaning-up operations in the areas affected by the spill, and in order to determine as far as possible the different levels of residual contamination, several sampling campaigns were carried out. The results of these sampling campaigns, with over 5000 samples taken, have been used as the basis for the study and analysis of the residual contamination and led us to define different intervention areas or plots where the measures to be undertaken would be different. Besides, the management of the collected data through the Environmental Information System will allow the elaboration of patterns for the assessment of contamination risks for species of interest, as well as predictive models of the flow and evolution of contaminants.

Concerning the soil contamination, once the clean-up finished and its state was known, remedial actions were designed to solve existing problems through the Guadiamar corridor. These were of the following types:

- physical removal of contaminated soil in those places where this was technically feasible
- chemical treatments to control pH in the soil, and as a consequence, reduce the solubility of the contaminants in such a way that they would remain in the soil and would not further contaminate the groundwater. This was mainly done by increasing the carbonate content
- control of metal ion mobility to immobilise the residual contaminants. Thus, additions of iron-rich clayey materials were made, mainly in areas where arsenic was an important problem
- phyto-remedies: these will be designed as the final operations in the remedial stage and will be limited to specific and controlled zones where physical or chemical treatments were not possible or not efficient. Tests are currently being carried out within the framework of agreements between the Regional Government of Andalusia with universities and the Spanish Scientific Research Council (CSIC)

Where residual levels were adequate for a later regeneration and reforestation, organic additions have been applied to improve soil properties and soil fertility, thereby favouring the planting and sowing of the vegetation. A long-term monitoring programme is currently being conducted and from the periodically obtained data it can be concluded that the current situation is good, being the same as the situation before the accident in many aspects (Ferrer *et al.* 2002).

15.4.3 Ecological restoration

A large part of the proposals emerging from the other strategy worklines justify the restoration or rehabilitation programmes. Here, the word restoration means the restitution of the structure and functioning of the aquatic and terrestrial ecosystems that form the Guadiamar river corridor and its watershed. It is an action process that aims to restore the ecological systems, as far as possible, to the dynamic condition they once had before being affected by a whole series of disturbances of anthropogenic origin, such as the effects of the mining spill and the long-term change in ecosystem structure.

The ecological restoration procedure (Prat *et al.* 2001) implies that the restored ecosystems should be self-maintaining in their functionality. For this reason, it is necessary to act on the key biophysical processes that determine their structure, functioning and dynamics. Under this approach, the restoration of the Guadiamar does not favour any particular element in the abiotic or biotic structures of the ecosystems, as could be the case for some emblematic species. Instead it focuses on the recuperation of the whole mesh of biophysical relationships that characterise their functioning and development over time. Moreover, the action programmes elaborated in the context of this workline take into consideration the great recuperation power of most of these aquatic and terrestrial systems, which appear to have great capacity to return to a benchmark or dynamic state after the end of a natural or human-induced temporary perturbation.

Therefore, implementation of 'passive' restoration procedures, implying the elimination of the most important factors that prevent or hamper their natural recuperation, has been considered essential. Next to that, active restoration programmes have also been included. In general, these programmes have been designed for the recuperation of the dynamic equilibrium within the fluvial system or its capacity to remain within the range of benchmark biophysical conditions. In this context, the priority actions to be taken were as follows.

Restoration of the geomorphologic system underlying the Guadiamar river corridor, based on its dynamic ecosystem functioning

For this purpose, recuperation of the stream dynamics (balance between the erosion/transport/sedimentation processes) was considered more important than to try to reconstruct its morphological structure. It is necessary, then, to leave the Guadiamar river valley as a *fluvial space* or a *free space*, through which the river might change its course without restrictions while recovering its longitudinal profile and acquiring its typical features of sinuosity, heterogeneity and connectivity.

*Restoration of the continuity of the Guadiamar river corridor as a linear
ecosystem that would process lateral, longitudinal and vertical flows of material,
energy and species*

For this reason, it is deemed essential to eliminate all the linear structures that
cut across its river bed, because they disrupt the continuity process.

Restoration of the natural perturbation regime

This is fundamental to maintain the ecological functionality of the river corridor. Its ecosystems do maintain their integrity by adapting their structure
and functioning to the intensity, magnitude and frequency of the yearly or seasonal natural perturbations, such as floods or droughts. These natural events,
which occur at different recurrent times, do not destabilise the floodplain or
river bed, but are a key ecological factor in the maintenance of corridor functionality. In this sense, and in order to ensure a quick passive restoration of the
river corridor, it requires the conservation of its natural flow rates, something
that is rather easy since the Guadiamar is one of the few Andalusian rivers that
can function without hydrologic control.

Restoration of vegetation communities

The functional restoration of the ecological corridor implies the restoration of
the original plant communities through the whole catchment independently
of its affectation by the spill, because the future Green Corridor should have a
diverse composition with respect to its habitats. Thus, with the aim of ascer-
taining which would be the different existing ecosystems if the anthropogenic
alterations had not happened, several studies have been planned. The difficul-
ties in reaching the planned goal were numerous, among other reasons because
seed banks have disappeared and areas with well-conserved vegetation are few
and small. A phytosociological study was undertaken and its output has been
combined, among others, with ecological variables, water table level, pedol-
ogy data, and data on its longitudinal and transverse location with respect to
the river, producing different vegetation models (Figure 15.6). In addition, owing to the importance of monitoring the behaviour of vegetation, an ambitious
plan has been designed and is currently being carried out. It tries to determine
in what degree the vegetation systems have a successful development according to the actions that have been undertaken. The conclusions will help us to
decide on the corrections that might be needed.

15.4.4 Integration of the human and natural systems of the watershed

It is evident that the capacity of the Green Corridor to attain its objectives would be substantially reduced if the human dimension of the Guadiamar
watershed had not been taken into consideration as an internal and dynamic component, rather than as an external and disruptive one. From this

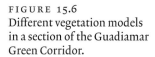

FIGURE 15.6
Different vegetation models
in a section of the Guadiamar
Green Corridor.

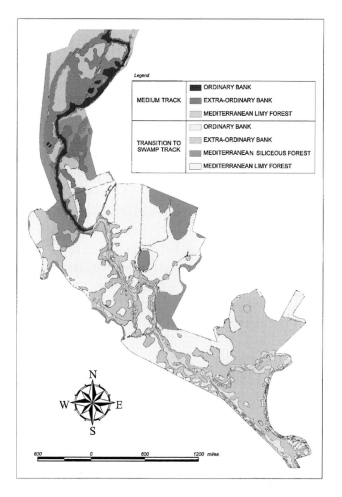

standpoint, it must be considered that the creation of the Guadiamar river
corridor as an ecological system must be socially valued in respect of the multiple assets (structural elements) and services (use of its functions) that can be
provided to the human systems. For this reason, the strategy developed a specific workline related to the integration of the human and natural systems.

Firstly, a sustainable development programme was furthered and promoted,
starting with identifying the most relevant sectors and activities in the conservation of the natural systems. Within each and every activity the *best
management practices* for ecosystem functionality are singled out. Subsequently,
we are researching and activating the measures and subsidies needed to advance the programme. Agriculture and livestock breeding activities developed
within the ecological limits of the river corridor, as well as leisure activities
and rural and nature tourism, are considered the priority sectors to be promoted. This is done through projects on sustainable agriculture and livestock
breeding financed through substantial funding from the European Union

(Regulation 2078/92, agri-environmental measures, and European Forestry Regulation 2080/92).

The other socio-economic sector of interest that must be promoted is the service of the ecosystem to society using natural systems for leisure activities. It is understood that the Guadiamar river corridor will fulfil important cultural, tourism and leisure activities that will produce substantial direct and indirect economic profits. For this reason, this sustainable development programme seeks to analyse multiple aspects such as the demands and needs for the use of natural resources by the local communities. An evaluation will be carried out on the environmental and landscape resources of the Guadiamar River and its watershed, the potential impact of tourism, leisure and sports activities on the ecosystems, its absorption capacity and the incorporation of the Guadiamar corridor into the significant tourism activity existent around Doñana.

The other two programmes that are included are oriented towards the notably important task of increasing the awareness of and communicating with the different socio-economic agents and institutions that are involved in the management of the natural systems of the watershed. The Environmental Education and Communication Programme endeavours to use the Green Corridor as a basic educational tool to increase the environmental awareness of the population. Moreover, the Public Use and Citizen Participation Programme intends to promote the active participation and commitment of the local and Andalusian communities in the implementation of the project.

15.5   PICOVER: a multi-disciplinary applied research programme

PICOVER (Box 15.1) is the most important, most ambitious multi-disciplinary applied research programme ever undertaken in Spain to address, from a scientific perspective, a social and environmental problem of a scale such as that caused by the Aznalcóllar mine spill. Scientists, the technical staff of the public administrations, and politicians are doing pioneering work together, sharing management and decision-taking responsibilities. Therefore PICOVER is a key component of the Guadiamar Green Corridor Project.

Despite all that has been achieved, there is still a long road ahead and many challenges to be met in order to fully carry out the task of creating the Guadiamar Green Corridor.

In the first year, necessarily, there was a great demand for information required to take decisions under the Urgent Measures Plan for minimising the impact of the Aznalcóllar mine spill. Of the four lines of research that make up PICOVER, the one related to the monitoring, control and correction of metallic contamination took precedence. For this reason, there is now a large amount of information and knowledge available about the consequences, scope and

BOX 15.1. Milestones in the development of the Guadiamar Green Corridor project

After three years' operation, the PICOVER Research Programme has been consolidated and the remaining elements that make up the Green Corridor Strategy assembled. During this time, intense, stimulating work has been done, whose major landmarks and achievements have been:

- the creation of a conceptual framework and methodological procedures that, based on the corpus of knowledge on the Management of Ecosystems, have allowed the development of the scientific basis for the Guadiamar Green Corridor Strategy or Action Plan
- an International Seminar on Ecological Corridors and the Restoration of Rivers and River Banks that has allowed discussion of the Action Plan in a multi-disciplinary forum, thus allowing its content to be enriched and its lines and programmes of action to be fine-tuned
- the creation of an open, flexible organisational structure, which allows the research agreements to be co-ordinated and PICOVER to be integrated into the system for decision-making on the design and execution of the lines and programmes of the Strategy
- the establishment of the PICOVER Co-ordination Commission, where the main researchers of the Research Projects, together with the managers, discuss and decide on the priority areas for research in accordance with the results emerging from the implementation of the Strategy
- the development of a Programme of Co-ordinated Monitoring of the activities involving the surveillance and control of contamination and the restoration of ecosystems degraded or destroyed by the mine spill, aimed at the recuperation of the Guadiamar river corridor

evolution of the effects of the mine spill on ecosystems and organisms, and this is allowing solid management decisions to be taken. The remaining lines of research have been under way since the beginning of the year 2000, in line with the established programme of work.

15.6   The future protection of this natural space

In our region the possibility exists of creating a network of ecological corridors based on the existing network of livestock routes and other elements to consider such as riverbeds, groves, basins, and public forests. An adequate management of these natural and cultural elements of our geography will

strengthen the linkage of the zones in the catchment and its objectives and reinforce the viability of the Nature 2000 network in Andalusia.

The management of such a diverse and special territory should be done in
an integral manner, giving a guarantee of spatial and normative measures that
permits the establishment of its ideal optimal functioning. The opportunity
presented is that almost the entire Green Corridor is in public ownership. This
greatly facilitates the management of this natural space, public use, services to
citizens, and environmental education with the conservation and recuperation
of its biodiversity. This management, combined with the local participation of
citizens, might need a specific denomination as already recognised by the laws
for natural spaces existing in Spain, both on a national and a regional level.

International categories, such as Biosphere Reserve, which the Doñana Park
already benefits from, have established criteria for degraded areas in representative zones. They should have unique natural characteristics and have the objective of restoration of the degraded ecosystem and augmenting in this way
the biodiversity in the area in question. It also can serve as a laboratory for ex-
perimenting with the techniques for recuperation and restoration of ecosystems to be able to apply them in other areas of the planet. It makes sense to do
this in the Green Corridor of the Guadiamar since it is a true space for experi-
mentation for the techniques of regeneration of marshland ecosystems and of
zones influenced by mining activities (Program MaB. UNESCO and PNUMA,
1974).

The recuperation of the degraded landscape presented in the Guadiamar
basin could be a sufficient condition for declaration of this space as a Protected
Landscape. This would have the objective of recuperation of its ecosystems and
its role of connecting two large natural Andalusian areas. The IUCN, in its
document of categories of natural space protection, establishes the category of
terrestrial and marine landscape. It considers that those spaces in which the
interactions of human beings and nature throughout the years has produced a
zone of definite character with important aesthetic, ecological, and/or cultural
values, and that normally shelter a rich biological diversity, merit this category.

In addition, Directive 92/43 of habitats of the European Union emphasises
the importance of this type of habitat for the survival of the Nature 2000
Network. It states that the member states will make an effort to foster the management of the linear elements of the landscape having a fundamental importance to the wild fauna and flora (articles 3.3 and 10). These are defined as those
elements which, for their linear and continuous structure (such as rivers with
their corresponding river beds or the traditional systems of demarcation in
the countryside) or for their role as a link point (such as ponds or groves), are
essential for migration, geographic distribution and genetic exchange of wild
species.

In Spain, Law 4/89 for the conservation of natural spaces and of the wild flora and fauna (Ministry of Agriculture 1989), determines that the protected landscape has a preferential treatment and similar importance is given to the other elements of flora, fauna, and minerals. The figure of the Protected Landscape is created for concrete places of the natural medium that, for their aesthetic and cultural value, merit special protection. These landscapes are designated by means of a law, and under these laws Peripheral Protection Zones could be established with the determination of avoiding ecological impact originating in the exterior. As it proceeds, the law itself will establish the necessary limitations.

15.7 Conclusions

The Guadiamar Green Corridor is an efficient instrument that can remedy some of the problems associated to the ecological fragmentation processes. Its implementation is based on the scientific knowledge of the natural environment and its components as well as the control of the results using the appropriate monitoring means available. Thus, the Guadiamar River Corridor gives a magnificent opportunity to reverse the recent historical trend in Doñana leading to the loss of its natural resources, and it can also be used to recover the natural dynamics of the ecological processes in south-west Spain.

The existence of a Pan-European Ecological Network is a stimulus for the programme, because it endorses the implementation of this corridor inasmuch as it comprises highly diversified biological and landscape areas. Likewise, the objective for the Natura 2000 Network, within the Andalusian context, is to connect the protected areas, whenever possible, in order to ensure the incorporation of the areas that would enhance the east–west connection both inside Sierra Morena as well as along the Doñana coast.

The ecological corridor of the Guadiamar will require a tailor-made protective body that, in due course, would ensure its functionality and sanction its inclusion in the RENPA. The elaboration of a Natural Resources Planning Scheme for the whole Guadiamar watershed is essential and it is being considered. This idea allows us to consider the possibility of defining and applying a new protective body. The possible environmental management body for the Guadiamar corridor must safeguard the participation of society in the decision-making process.

16

Conclusions: into the twenty-first century

Before the path existed, one always had to be in an automobile to get anywhere in the community. Now people meet on the path to walk side by side and talk. The path has given the town something to be united about, which has cut across all brackets of income, age, place of residence, and at the same time has preserved the spectacular, previously private riverbank for all to enjoy all seasons of the year.

(*A resident of Stowe, Vermont; Schwarz 1993*)

16.1 The context

Developments in science and in the application of scientific principles are progressing fast. Land use change is progressing even faster. We cannot look far into this new century; still we can detect some important trends and issues. The design of ecological networks is based on concepts, within certain scientific and planning contexts mediating specific values, traditions and relations of power. Some concepts are selective and will only be valid under the particular circumstances in which they were created. Elaborate ideas of a national ecological network make little sense in a big country such as Russia, and would not within the time limits of one generation be given the large amounts of funding that are needed. The more comprehensive ecological networks of the relatively sparsely populated Estonia and Lithuania will be easier to implement than would be the case in Denmark, the Netherlands or Poland. Strategies will also differ between New England and Florida on the one hand, and Argentina and Northern California on the other hand. Although comprehensive networks make sense at the national level, not all countries follow this direction or support it politically.

Ecological connectivity, as pointed out by Ahern in chapter 3, is a prerequisite for the sustainability of certain landscape processes, such as animal and human movements, flux of nutrient and flow of water. The American concept of 'nearby nature' comes from the need to experience the natural world as a normality in everyday life, and has been shown to improve personal and social health. It is therefore a psychological and social benefit at the same time. The hypothesis of co-occurrence points out that greenways constitute an efficient and strategic method to protect natural resources with the least amount of land. Indeed environmental continuity supports landscapes that are multi-functional, and thus where multiple compatible uses can be managed. These landscapes become in this way the platform for a wider political support, which in turn allows longer protection of nature.

Lastly, the approach of combining recreation and nature conservation is becoming more common and, at least in the USA, has already replaced centralised parks and recreational planning (Fabos 1995). In Europe econets are included in the legislative and planning systems of several national and regional governments (Jongman 1995), and other actions are going to be taken into account.

16.2   A framework of linkages

The linkage approach constitutes a foundation for landscape development at different scales (Forman 1995), especially at the regional level. Corridors intersect a landscape to form a network of core areas. Networks are different in purpose, function and domain. There are single-, dual- and multi-purpose networks. They can function at ecological, human and artificial levels and they can have ecological, hydrological, anatomic, infrastructural, agronomic, pastoral, recreational, cultural, communicational and informatic domains.

Both econets and greenways constitute a triple framework. Firstly they provide a conceptual framework in landscape ecology, nature conservation and recreational activities. Secondly they create a framework of different domains, including human, on the Earth. Thirdly they form an ecological framework of ecosystems and habitats in the landscape. The last case can certainly be used as a framework for planning and policy actions, able to enhance the ecological infrastructure and thus to conserve the biological and landscape diversity of a region. Moreover, it can assist in revisiting other areas that have implications for nature conservation (Bennett 1994).

Yet econets and greenways have recently moved fast from a conceptual to a design stage, both in the planning environment and *in situ*. Networks of different types are designed, scenarios are prepared, connectivity is restored, and nature is protected and also re-created. Nevertheless, to clarify which parts of an area should be included in the ecological network we need to ask the following questions: what, where and how. This has to be done on different levels: from the estate and community levels, to regional, national and international ones. What is the need, what will potentially be saved and what will not be saved in the strategy pursued? What is the optimal network and where should it be located? How will it function or dysfunction, and last but not least how are we going to implement these ecological networks? Most of them are in a developing phase, and the chapters in this book indicate that there is enthusiasm about the idea of ecological networks among conservationists, planners, NGOs and politicians in different locations, regions and whole countries. However, no single theory, no single scientific or planning concept can be taken as universally applicable. Our available knowledge is not sufficient to predict in detail what the advantages and disadvantages are.

In many countries there is a link between development of ecological corridors and cultural landscapes. The possibilities that are provided to establish and protect landscape features on farmland in all of the selected countries are good examples. In Italy, but also in Denmark and Estonia, emphasis lies on the link with cultural landscapes. In Denmark the concept of ecological corridors is linked to considerations of outdoor life, and there has been an important recreational aspect in creating ecological corridors. Protection of landscape elements such as stone and earth dikes has a meaning for ecological corridor protection, as well as a cultural and historical value. In the Netherlands the link with cultural landscapes also exists. Many nature areas of the national ecological network are open to recreational use and this may also apply to ecological corridors. In the ecological network of Slovakia, the terrestrial system for ecological stability (TSES) emphasises the fact that it is being developed for the whole country on different scales, with the local scale including cultural landscapes. The development of TSES and corridors is mainly for the sake of preservation of biological and landscape diversity, but also to connect the natural landscape with the land used by man for settlements.

The need for corridors at the continental scale is part of the American Wildlands Project (Dobson *et al.* 1998). It has also been discussed by Bouwma *et al.* in chapter 6, where a methodology to assess this need has been proposed, including an indicative and simplified analysis of target species for European corridors. The limitation arising in working at large scales has been pointed out, as well as the necessity of more detailed analysis.

16.3   Planning and implementation

In Europe ecological networks and greenways show a fast development: from a few networks at the end of the 1980s to over forty now. In the Americas and in Australia the same developments can also be seen (Bennett and Wit 2001). Implementation is starting in a number of countries. The Slovak, the Czech and the Dutch approaches are similar in the sense that they are elaborated at four different hierarchical levels: the national, regional, municipal, and cross-border. The planning system is a centralised top-down one, agreed upon by national governments. Nevertheless, emphasis is now being given to regional development of ecological networks.

Ecoregional planning can be employed on a larger scale of regional nature conservation, as for the Florida Ecological Network illustrated by Smith in chapter 5. In this case the design of the network has been considered an iterative and dynamic process, continuously enriched by new data, information and additional species-specific habitat analysis. Legislative actions and acquisition programmes can support the regional planning process, while rural stewardship programmes can support the ecological network implementation. To avoid conflicts between land use and nature conservation, ecological network development can be used as a planning tool to assist in the strategic selection of present expenditures (e.g. land acquisition and conservation) and future gains (e.g. natural resources and quality of life).

The fortunate situation of the Florida Ecological Network, where it is possible to protect large intact landscapes within an ecological network, allows moving from the advantage of merely the present population to the benefits of future generations, leaving to them a good quality of biodiversity and natural heritage. This ideal picture can, however, come about only with the expertise of scientists and planners, the willingness of politicians and decision-makers, and the support of local government and above all the local people.

Nearly all plans that have now been developed in Europe (Jongman and Kristiansen 2001) have regional aspects and include instruments at the regional level. Most show linkages with the local level where implementation has to take place. In this context, regional and local have to be considered as a policy level, not as a spatial level. Regional is linked with regional authorities between national and local, where local is the level of municipalities. Several chapters in this book show the importance of the regional and local level in the implementation. The design for the provinces of Bologna, Modena and Gelderland (chapter 8) as well as the approaches in Argentina (chapter 14) for involving local people are emphasising the need for it.

In southern Europe, national planning has at times faced problems and therefore in some places planning has been based on a more local and regional level. This is illustrated by the case of Italy, where the provinces are increasingly active in developing ecological networks as a concept for landscape planning (Cavalchi and Pungetti 2000), and by the case of Spain, where ecological networks and corridors have developed at regional and local level. In Portugal the national network of protected areas does already exist and local pilot projects are being worked out. The same situation holds in other parts of Europe, as in Ireland, where a national pilot project is being set up. All these approaches, however, have in common that they act where the need is. In other countries a unified national approach will be hard to achieve and maybe it would not make any practical sense. Russia in particular has such a huge and partly unexplored diversity of both natural and cultural features that a national approach would be unachievable within a limited time period.

The wider landscape is the appropriate context in the ecological network development of areas of ancient human usage. However, as suggested by Pungetti and Romano in chapter 7, the development can be successful only if supported by applied landscape ecology, careful environmental policy and knowledge of the local cultural landscape in terms of its evolution, values, tradition and use. The wider public context should consequently be taken into account in that development. Regional and national policies, for example, can assist the European legislation in the comprehension of local ecological structures, socio-economic needs and pressures on the regions (Pungetti 1991). Accordingly, it is proposed that ecological network development in Europe will integrate natural and cultural aspects. This means to consider the socio-economic trends and functions of the site, linking ecology to environmental economics, and then to integrate ecological networks into the planning system in a holistic perspective of natural, cultural, social, economic and political considerations.

The planning and management of nature conservation have to be based on a decentralised framework, especially if nature outside reserves and national parks is to be included. Spain, Italy, Austria, Denmark and Germany are examples in Europe of mainly decentralised approaches based on formal divisions of tasks between the national state and the regions or Länder. In the USA a comparable division of tasks can be seen, giving a major responsibility to the states and to those federal organisations that are decentralised. In a number of countries a decentralisation process is ongoing. The consequence of the decentralised character of nature conservation is that international co-operation involves not only countries, but also autonomous regions, Länder, counties, districts, provinces, cantons and amte, or whatever the level and the name of the responsible region.

Greenways in the USA, as well as ecological corridors in Europe, can be linked to larger networks, using landscape planning at a regional scale. Furthermore, implementation tools such as land acquisition, regulations and stakeholder involvement can be applied. This type of implementation allows the move from a visionary to a realistic perspective, closer to the requirements of both nature and society. Greenway implementation indeed involves not only governmental agencies, but also NGOs and the general public. This particular perspective, as illustrated by Erickson in chapter 12, has gained wider popularity and has therefore supported the consideration of greenways in the planning of most large cities of the USA. Owing to their characteristic of linear open spaces along rivers, canals, rail lines and roads, they intersect both public and private lands and a variety of land uses.

Whereas in Europe the most relevant objective in ecological network development has been nature conservation and restoration, in North America common objectives are also recreation, neighbourhood enhancement and, recently, environmental education. In such a complex system, it is clearly important to set up broad objectives that can be directly applied in the planning system where greenways or econets are developed. Long-term plans, aimed to integrate land use, transport, water and open spaces, should therefore be envisaged, and fostered by public and private economic development.

In both Europe and North America, however, the proposal of new objectives in nature conservation and land use planning encounters not only physical barriers but also cultural ones. Among the latter are cultural conformity, bureaucratic obstacles, political decisions, fear of the unknown, lack of vision, and last but not least social and ecological illiteracy (see Figure 7.9). The setting up of clear objectives, consideration of local needs, direct involvement, education and greater awareness could help in this dilemma.

16.4   Land use implications

Land use is an essential element to consider in the development of ecological networks and greenways. Ecological networks with small-scale linkages and barrier passages are less urgent in regions with low land use pressure such as Siberia, parts of the western USA and western Ireland. However, even in an area with relatively low population density, such as Jujuy in Argentina, the problem of linkage does occur, as it does also in the western part of the USA and in Siberia. The need for landscape linkages or ecological corridors is related not only to the absence of land use pressure, but also to the kind of species that are involved. The Siberian tiger and the Argentinian jaguar require wider undisturbed habitat areas and ecological corridors than does the relatively poor island fauna of Ireland.

But we also have to look forward. In some regions of central Europe large areas of land are intensively used for agricultural production. In the future these lands will be able to compete in food production with the rest of Europe and the USA, and therefore intensification of land use can be expected here too. Problems for ecological linkages should be solved in the period of accession to the European Union.

In western Europe the possibility to create ecological networks by the temporary marginalisation process has not been utilised owing to resistance from agricultural lobbies. In Denmark, for example, a consensus has existed between the agricultural and environmental ministries that agriculture in the temporary marginalisation process should not give up land for the purposes of conservation, but hold it to extend production in case of more prosperous times for agriculture in the future. Only lately have ecological networks been based on a more proactive strategy such as doubling the area of forests.

Looking at ecological networks as integrated systems of corridors and core areas, it is possible to see them as 'green infrastructures' to maintain or restore ecological processes and functions, and therefore to restore nature. They are, however, threatened by particular land uses such as infrastructure. Road networks are undeniably a prerequisite of present economic development, but human movement is often in conflict with wildlife movement. Many studies have been carried out on the impact of heavy traffic roads on core areas and other natural zones. Jaarsma in chapter 10 has complemented these with analysis of the conflict emerging when ecological corridors cross minor roads. Here the aim is to reduce and possibly avoid this conflict. A proposed solution is to balance human and wildlife movements within transport planning at a regional scale.

Wildlife crossings already exist in many places all over the world, in situations for example of high motorways, viaducts, railways and road tunnels. In the attempt to mitigate the barrier effects of transport routes on wildlife movement, road projects can be programmed and existing structures can be identified. In addition, mitigation measures, e.g. wildlife underpasses, can be planned and later implemented. However, a mitigation strategy should deal not only with mitigation measures, but also with land use control and conservation designations. On this line, purchase of land, zoning and land use control can be seen as unpopular restrictions among the population. To overcome this, alternative strategies can be proposed, such as control of partial interests of land or, as Smith suggests in chapter 5, conservation easements through donated, transferred or purchased development rights.

16.5   Partners and public involvement

The involvement of the public and the support of NGOs are prerequisites for a successful implementation of ecological networks and greenways. To our knowledge there are different historical preconditions for public involvement between countries, based on their context. For example, in some countries creating public access to nature is essential, whereas others are focusing on nature conservation. In Russia in the past there was the objective to exclude the public from the *Zapovednik* (National Park). This is now changing with a proper establishment of national parks and the yearly public arrangements of 'March for Parks' (Sobolev, pers. comm.). As shown by Jongman in chapter 2, in the UK the definition of national park explicitly mentions recreational use, whereas in both the IUCN and the German definition, strict protection has been seen as the most important feature of national parks. However, ecological network implementation means inclusion of nature and species movement outside protected areas. Therefore it implies that the need to secure public involvement and support for its development is a priority for the future.

The Danish counties have experienced difficulty in co-operating with actors involved at the local level. In Flanders the failure to present the value of the ecological network has led to strong political opposition in the phase of implementation (De Blust *et al.* 1995). That the development of greenways and ecological networks cannot happen without public involvement should be an important lesson for the new century. The more successful we are in winning public support, the more results we will gain.

Local NGOs and grassroots groups are important everywhere for the implementation of ecological networks and greenways. They play a fundamental role when nature protection legislation is violated, as both guardians and actors who care for nature. They are also important promoters of knowledge and mediators between expert and lay discourses, as well as between the national level and local communities. The American greenway movement is mainly a grassroots-based initiative. In Eastern Europe NGOs play the role of legitimate and effective partners on national, regional and local levels, thus supplying what is lacking in both nature conservation and democratic institutions. In some regions NGOs, together with some scientific bodies, are fulfilling functions that in other parts of the world are taken in hand by long-established institutions.

In the past decade a stronger sense of co-operation than ever before in nature conservation has developed all over the world. This has opened up the way to constructive discussion and easier flow of information. Several initiatives have been put forward under the name of greenways in the Americas and econets in Europe, including strategies, policies and planning actions by governmental and local bodies, environmental organisations, working groups and expertise networks. Paradoxically, however, while all these institutions emphasise action on paper (Bennett 1994), implementation on the ground is still difficult to achieve. An answer to this can be found in multi-disciplinary approaches, cross-boundary co-operation, cross-organisational partnership and information exchange between regional organisations and authorities. Without doubt the conservation of the world's natural heritage is a joint responsibility that requires co-operation, and an active and honest partnership.

It is also true that such a partnership can be easier or more difficult to establish in different situations and places. For example the interaction between natural and cultural systems can be very complex, as in Italy and Spain, owing to long historical activities and strong human presence. This creates a hurdle in the separation of natural and cultural processes, and consequently in the understanding of ecological issues. It has also implications in the cultural perception of the problem, often manifested in lack of environmental knowledge and awareness. Furthermore, there is sometimes hostility towards nature conservation in those populations interpreting it more as a present restriction than as a future advantage (Pungetti 2001).

One of the problems is that the ecological–scientific community and nature conservation authorities have often overlooked the human side of the coin in terms of perception, tradition, economy, society and even the needs of the local community, so often excluded from decisions on 'their' land development or conservation. An alternative approach is necessary in both communities, which should come closer, find common grounds and finally start effectively working together. From agreements on partnership, it is then possible to move to discussion and planning based on a multi-disciplinary approach, which is recently becoming more common as is shown in the approaches for Argentina (chapter 14) and Andalusia (chapter 15). A similar direction has been undertaken, for example, in France with the institution of interdisciplinary teams of landscape ecologists, agronomists, geographers and social scientists. Such teams can analyse not only ecological and landscape functions, but also the perception of greenways by the local farmers, by comparing several network structures in the same ecological–cultural region (Burel and Baudry 1995).

16.6   Instruments for ecological network and greenway development

Ecological networks and greenways are gaining growing attention. Financial instruments, such as tax support, are common to protect linear features and landscape elements. In addition to the instruments that already exist in several countries to protect landscape, instruments to protect ecological corridors are under development or revision in many countries. Spatial planning is the most important framework for the establishment of ecological corridors, and therefore nature conservation and land use instruments are the key tools to be used for their implementation.

In Europe there is a variety of approaches. In Flanders three laws provide for the protection of linear features at national level, while at local level the development plans of municipalities include management and protection of roadside verges and hedgerows. In Lithuania several possibilities for tax reduction can contribute to the conservation of ecological corridors. Also in Slovakia this instrument is used: tax reductions are available to farmers who use certain areas non-commercially. In Slovakia, in particular, there are several examples of protection of linear features: provisions are made for elements of landscape that contribute to its ecological stability, and farmers can be compensated for property losses that are the result of prohibitions or other measures of nature and landscape conservation. In Denmark protection of landscape elements without the specific aim of creating ecological corridors has traditionally developed in the schemes for the conservation of small biotopes, such as ponds and lakes, and other landscape features, such as stone and earth dikes, and for the establishment of hedgerows. Restrictions on land use, for example in coastal areas, and alteration of the buffer zones along watercourses and the edges of forests and lakes, can also contribute to the protection of ecological corridors.

In the Netherlands, the UK, Slovakia and Denmark schemes for Environmentally Sensitive Areas (ESA schemes) play a role in the protection of linear features of farmlands. This includes hedgerow planting, restoration of ponds, and the use of environmentally friendly farming methods. In Slovakia, specially protected areas of nature and landscape can be preserved by agreements between a nature protection body and the owner of the land with respect to its management. In Hungary and Estonia too, ESA schemes are being developed. In Estonia they will include the protection, among other things, of wildlife habitats and such landscape elements as farmlands and field boundaries. In Hungary the ESA scheme will include support for maintaining and restoring ecological corridors. Co-operation with farmers is given special attention because they are key actors in management. Studies of some European countries on agri-environmental measures illustrate the importance and possibilities of initiatives to protect landscape features on farmland by using financial instruments such as ESA schemes.

The connection between infrastructure and the environment poses further questions. An integrated regional planning approach can help in addressing the problem. Jaarsma's study in chapter 10 shows that if mitigating measures such as fencing and tunnelling are not feasible on rural roads, other technical solutions instead might help to improve wildlife road crossing. Among these is the tool of rural traffic calming, which also brings benefits to local people by increasing traffic safety and improving the quality of life in rural areas. In addition, the inclusion of mitigation and compensation measures in infrastructure projects, as illustrated by Smith (chapter 5) and Jaarsma, outlines that fragmentation problems should be tackled at different scale levels (international, national, regional and local). The actual realisation of the fauna passover at a local level is carried out on a number of general principles and within local or national specific circumstances.

Finally, multi-objective plans, as pointed out by Erickson in chapter 12, can be one option for future ecological network and greenway development, where the restoration of nature can proceed parallel to the restoration of communities. Tools such as complementary policies can be used, as well as incentives for biodiversity conservation, road amelioration and urban revitalisation. However, the *sine qua non* of greenway and ecological network implementation on the ground is a partnership of local agencies and stakeholders. Some of them participate in only a part of the project, but all are involved somehow. Facilitators and networkers play a crucial role, state or regional agencies participate to a greater or lesser extent in the founding or planning phase, NGOs provide connections and support from the communities, and the private sector can also support funding and strengthen environmental protection, bringing in community rights to be heard.

We like to think, like Erickson, that the process of creating ecological and cultural connectivity has neither a beginning nor an end. Like landscape, it is an evolutionary process varying in time and space, and therefore encountering problems but also evolving when opportunities arise.

16.7   Epilogue

In the first century BC, Lucretius illustrated that the survival of species is a responsibility that people have had since early times (Lucretius 1937). Of those species that existed at the period of Lucretius, a great number are now extinct or confined to isolated spots of the world. Almost two millennia later the same situation was again examined by Darwin, looking at evolution as a means for the adaptation of species to the natural and anthropogenic changes in the world (Darwin 1909). The order given to nature is nowadays created mainly by humans and their culture, and rests on the same principle of responsibility evidenced by both Lucretius and Darwin. On the basis of such a responsibility, action to preserve biological, landscape and cultural diversity is urgently required, and alternative solutions to balance natural and human development must be found.

From the chapters in this book it is possible to recognise how ecological networks and greenways do represent a solution. They are already a popular international movement and their success lies in the capability to go beyond the ecological perspective, embracing also social and political imperatives. They are in fact not only ecological networks, but also social and political networks able to connect different people. On account of this, econets and greenways are an alternative approach to strategic planning in driving environmental and political factors, where links and connectivity play a key role. Strategic planning nevertheless has to be flexible, in order to adapt to local circumstances that continuously change over time. Planning and management decisions can therefore be re-conceived and re-drawn in the view of an adaptive approach. Moreover, strategic planning cannot stand alone: it has to be implemented through a collaborative partnership of selected bodies and stakeholders.

The case studies in this book illustrate common aspects and different circumstances where econets and greenways are now being planned and developed, and also show ways to deal with the matter. Undoubtedly it will take time to achieve effective species and habitat protection. New legislation has to be passed, new plans drawn and new management taken on. But the hardest task is to build partnership, co-operation and trust among people. The state of the art is, however, positive, considering the rapid development of econet and greenway concepts in American and European regions.

For ecological network and greenway implementation, it is clearly necessary to develop design techniques, bridge disciplines, push different public sectors to work together, gain support, and above all engage stakeholders and co-operate with local communities. This is a way forward to a world no longer separated, fragmented and degraded, but based on environmental continuity, linkages, co-operation, responsibility, health and enjoyment. Do not take anything for granted: experiment, and weigh up the experience of others like that above of the resident of Stowe, Vermont.

References

Ahern, J. 1991. Planning for an extensive open space system: linking landscape structure and function. *Landscape and Urban Planning*, **21**, 131–45.

1994. Greenways as Ecological Networks in Rural Areas. In E. A. Cook and H. N. van Lier (eds.), *Landscape Planning and Ecological Networks*, pp. 159–177. Elsevier, Amsterdam.

1995. Greenways as a planning strategy. *Landscape and Urban Planning*, **33**, 131–56.

1996. Greenways as a Planning Strategy. In J. Fabos and J. Ahern (eds.), *Greenways: the Beginning of an International Movement*, pp. 131–55. Elsevier, Amsterdam.

Ahern, J. and Fabos, J. 1998. A Global Greenway Vision: The Role of GIS. GisPlanet Conference, Lisbon, Portugal. *CD-ROM Proceedings GIS PlaNET'98.*

Andrews, A. 1990. Fragmentation of habitat by roads and utility corridors: a review. *Australian Zoologist*, **23**(3,4), 130–41.

APA (American Planning Association) 1995. *Environment and Natural Resources*. Capitol Highlights, A publication of the Florida Chapter of the American Planning Association. Issue no. 95-1.

Arambarri, P., Cabrera, F. and Toca, C. 1984. *La contaminación del río Guadiamar y su zona de influencia, Marismas del Guadalquivir y Coto de Doñana, por residuos de industrias mineras y agrícolas.* CSIC, Madrid.

Arrieta, J. and Pastor, C. 2000. *Relevamiento Socioeconómico Ambiental de las Comunidades del Tramo Medio e Inferior de la Cuenca del Río Bermejo.* In Informe Final del Programa Estratégico de Acción para la Cuenca del Río Bermejo, OEA, Buenos Aires.

Arts, G. H. P., van Buuren, M., Jongman, R. H. G., Nowicki, P., Wascher D. M. and Hoek, I. H. S. 1995. Editorial. *Landschap* (Special issue on ecological networks) **12**(3), 5–9.

Associated Press. June 20, 2001. Land Conservation Bill Moving Ahead. *New York Times.*

Bani, L., Baietto, M., Bottoni, L. and Massa, R. (2002). The use of focal species to design a habitat network for a lowland area of Lombardy, Italy. *Conservation Biology*, **16**(3), 826–31.

Bardach, E. 1977. *The Implementation Game: What Happens After a Bill Becomes a Law.* MIT Press, Cambridge, MA.

Baschak, L. and Brown, R. 1994. River systems and landscape networks. In E. A. Cook and H. N. van Lier (eds.), *Landscape Planning and Ecological Networks*, pp. 179–99. Elsevier, Amsterdam.

Battisti, C. and Contoli, L. 1997. Sulla componente di ricchezza della diversità avifaunistica in Italia: peninsularità ed insularità. *Riv. Ital. Ornitologia*, **67**, 113–26.

Baudry, J. and Merriam, H. G. 1988. Connectivity and connectedness: functional versus structural patterns in landscapes. In K.-F. Schreiber (ed.), *Connectivity in Landscape Ecology. Proceedings of the 2nd International Seminar of the*

International Association for Landscape Ecology, pp. 23–28. *Münsterische Geographische Arbeiten* 29. Münster.

BEBR (Bureau of Economic and Business Research) 1998. *Florida statistical abstracts*. University of Florida, Gainesville, FL.

Behn, R. D. 1988. Management by Groping Along. *Journal of Policy Analysis and Management*, **7**, 643–63.

Beier, P. 1995. Dispersal of juvenile cougars in fragmented habitat. *Journal of Wildlife Management*, **59**, 228–37.

   1996. Metapopulation models, tenacious tracking, and cougar conservation. In D. R. McCullough (ed.), *Metapopulations and Wildlife Conservation*, pp. 293–324. Island Press, Washington, DC.

Beier, P. and Noss, R. F. 1998. Do habitat corridors provide connectivity? *Conservation Biology*, **12**, 1241–52.

Bellamy, D. J. 1994. Windows of Hope: a sideways look at the biological diversity of Europe. In G. Bennett (ed.), *Conserving Europe's Natural Heritage: Towards a European Ecological Network*, pp. 35–40. Graham & Trotman, London.

Bemis, W. E. and Kynard, B. 1997. Sturgeon rivers: an introduction to acipenseriform biogeography and life history. *Environmental Biology of Fishes*, **48**, 167–83.

Bennett, A. F. 1999. *Linkages in the Landscape: The role of corridors and connectivity in wildlife conservation*. IUCN, Gland.

Bennett, G. (ed.) 1991. *Towards a European Ecological Network*. IEEP, Arnhem.

   (ed.) 1994a. *Conserving Europe's Natural Heritage: Towards a European Ecological Network*. Graham & Trotman / Martinus Nijhoff, London.

   1994b. A European approach to nature conservation. In G. Bennett (ed.), *Conserving Europe's Natural Heritage: Towards a European Ecological Network*, pp. 43–9. Graham & Trotman / Martinus Nijhoff, London.

   1998. *Guidelines for the Development of the Pan-European Ecological Network*. Draft. Council of Europe, Committee of Experts for the European Ecological Network. STRA-REP (98). 6.

Bennett, G. and Wit, P. 2001. *The development and application of ecological networks. A review of proposals, plans and programmes*. AIDenvironment, Amsterdam.

Berthold, P. 1993. *Bird Migration. A General Survey*. Oxford Ornithology Series No. 3. Oxford University Press, Oxford and London.

Bischoff, N. T. and Jongman, R. H. G. 1993. *Development of Rural Areas in Europe: The Claim for Nature*. Preliminary and background studies no. V79. Netherlands Scientific Council for Government Policy. SDU Publishers, The Hague.

Bohemen, H. D. van 1995. Mitigation and compensation of habitat fragmentation caused by roads: strategy, objectives and practical measures. *Transportation Research Record*, **1475**, 133–7.

Bohemen, H. D. van, Buizer, D. A. G. and Littel, A. (eds.) 1991. *Nature Engineering and Civil Engineering Works*. Pudoc, Wageningen.

Boitani, L. 1997. Il ritorno dell'Orso. *Attenzione, WWF Italia*, **7–8**, 11–16.

Boscagli, G. 1988. *L'Orso*. Lorenzini, Udine.

   1985. *Il Lupo*. Lorenzini, Udine.

Bouwma, I., Jongman, R. H. G. and Butovsky, R. O. (eds.) 2002. *The indicative map of the Pan-European Ecological Network for Central and Eastern Europe*. Draft version. ECNC, Tilburg.

Boyne, G. A. 1992. Local government structure and performance: Lessons from America? *Public Administration*, **70**, 333–57.

Brandt, J. 1995. Ecological networks in Danish planning. *Landschap*, **12**(3), 63–76.

Bridgewater, P. 1988. Ecolines and geolines: connectivity in natural landscapes. In K.-F. Schreiber (ed.) *Connectivity in Landscape Ecology. Proceedings of the 2nd International Seminar of the International Association for Landscape Ecology. Münsterische Geographische Arbeiten* 29. Münster.

Broekhuizen, S. 1986. De betekenis van kleine landschapselementen voor marterachtigen. In P. Opdam, T. A. W. van Rossum & T. G. Coenen (eds.), *Ecologie van kleine landschapselementen*, pp. 45–52. RIN, Leersum.

Brooker, L., Brooker, M. and Cale, P. 1999. Animal dispersal in fragmented habitat: measuring habitat connectivity, corridor use and dispersal mortality. *Conservation Ecology* (online), **3**(1), 4.

Bryant, R. 1998. Bigger is better. *Urban Land*, **57**(7), 60–2, 85–7.

Buček, A., Lacina, J. and Michal, I. 1996. An ecological network in the Czech Republic. *Veronica*, Special 11th Issue, 1–44.

Buchwald, K. and Engelhardt, W. 1980. *Handbuch für Planung, Gestaltung und Schutz der Umwelt*. BLV Verlagsgesellschaft, Munich.

Bunce, R. G. H., Smart, S. M., Van de Poll, H. M., Watkins, J. W. and Scott, W. A. 1999. Measuring change in British vegetation. *Ecofact* Volume 2. ITE, Merlewood Research Station.

Burel, F. and Baudry, J. 1995. Social, aesthetic and ecological aspects of hedgerow in rural landscapes as a framework for greenways. *Landscape and Urban Planning*, **33**, 327–40.

Burkhardt, R., Jaeger, U., Mirbach, E., Rothenburger, A. and Schwab, G. 1996. Planning habitat network in Rheinland-Pfalz, Germany. In P. Nowicki, G. Bennett, D. Middleton, S. Rientjes and R. Wolters (eds.), *Perspectives on Ecological Networks*, pp. 19–29. ECNC, Tilburg.

Cao, H., Francés, M. and Vaca, A. 1997. Empleo público en las provincias rezagadas. In *Revista APORTES para el Estado y la Administración Gubernamental*, No. 8. Buenos Aires, Asociación de Administradores Gubernamentales. (www.ag.org.ar/aportes.htm n8)

Cates, C. 1979. Beyond muddling: creativity. *Public Administration Review*, **39**, 527–32.

Caughley, G. and Sinclair, A. R. E. 1994. *Wildlife Ecology and Management*. Blackwell Scientific Publications, Boston, MA.

Cavalchi, B. and Pungetti, G. (eds.) 2000. *Verso la realizzazione di reti ecologiche in aree rurali*. ARPA, Bologna.

Cisneros, H. G. 1995. *Regionalism: The New Geography of Opportunity*. US Department of Housing and Urban Development, Washington, DC.

Clevenger, A. P. and Waltho, N. 2000. Factors influencing the effectiveness of wildlife underpasses in Banff National Park, Alberta, Canada. *Conservation Biology*, **14**(1), 47–56.

Clobert, J., Danchin, E., Dhondt, A. A. and Nichols, J. D. (eds.). 2001. *Dispersal*. Oxford University Press, Oxford.

Consejería de Medio Ambiente 2000. *Programa de Investigación del Corredor Verde del Guadiamar-PICOVER*. Junta de Andalucía, Sevilla.

Cook, E. A. and van Lier, H. N. 1994. *Landscape Planning and Ecological Networks*. Elsevier, Amsterdam.

CORINE 1991. *Biotopes Manual, Habitats of the European Community*. EUR 12587/3. Office for Official Publications of the European Communities.

Council of Europe, UNEP, and European Centre for Nature Conservation 1996. *The Pan-European Biological and Landscape Diversity Strategy, a Vision for Europe's Natural Heritage*. Council of Europe, UNEP, ECHC.

Cowardin, L. M., Carter, V., Golet, F. C. and LaRoe, E. T. 1979. *Classification of Wetlands and Deepwater Habitats of the United States*. US Fish and Wildlife Service, Washington, DC.

Cowling, R. 1999. *Planning for persistence – systematic reserve design in southern Africa's Succulent Karoo Desert*. In *PARKS*, Vol. 9, No. 1, February 1999. Protected Area Programme. IUCN Gland.

Cox, J., Kautz, R., MacLaughlin, M. and Gilbert, T. 1994. *Closing the gaps in Florida's wildlife habitat conservation system: recommendations to meet minimum conservation goals for declining wildlife species and rare plant and animal communities*. Florida Game and Fresh Water Fish Commission, Tallahassee, FL.

Cristoffer, C. 1991. Road mortality of northern Florida vertebrates. *Florida Scientist*, **54**(2), 65–8.

Crooks, K. R. and Soulé, M. E. 1999. Mesopredator release and avifaunal extinctions in a fragmented system. *Nature*, **400**, 563–6.

Csuti, B. 1991. Conservation corridors: countering habitat fragmentation, an introduction. In W. E. Hudson, (ed.), *Landscape Linkages and Biodiversity*, pp. 81–90. Defenders of Wildlife, Island Press, Washington, DC.

Curatolo, J. and Murphy, S. 1986. The effects of pipelines, roads, and traffic on the movements of caribou (*Rangifer tarandus*). *Canadian Field Naturalist*, **100**(2), 218–24.

Dale, V. H., Brown, S., Haeuber, R. A., Hobbs, N. T., Huntly, N., Naiman, R. J. *et al.* 1999. *Ecological principles and guidelines for managing the use of land: A report from the Ecological Society of America*. (http:// esa.sdsc.edu/esalanduse.htm)

——— 2000. ESA Report. Ecological principles and guidelines for managing the use of land. *Ecological Applications*, **10**, 639–70.

Daniels, T. and Bowers, D. 1997. *Holding our Ground: Protecting America's Farms and Farmlands*. Island Press, Washington, DC.

Darveau, M., Beauchesne, P., Belanger, L., Huot, J. and Larue, P. 1995. Riparian forest strips as habitat for breeding birds in boreal forest. *Journal of Wildlife Management*, **59**(1), 67–78.

Darwin, F. (ed.) 1909. *The Foundations of the Origin of Species: two essays written in 1842 and 1844*. Cambridge University Press, Cambridge, UK.

Davies, C. E. and Moss, D. 1999. European Environmental Agency. European Topic Centre on Nature Conservation 1999 Work Programme: Task 4.3. *EUNIS Habitat Classification*. Draft Final Report, July 1999. ITE, Monks Wood, Huntingdon.

Davis, F., Stoms, D. and Andelman, S. 1999. Systematic reserve selection in the USA: an example from the Columbia Plateau ecoregion. In *PARKS*, Vol. 9 No. 1, February 1999. Protected Area Programme, IUCN, Gland.

Dawson, K. 1996. A Comprehensive Conservation Strategy for Georgia's Greenways. In J. Fabos and J. Ahern (eds.), *Greenways: the Beginning of an International Movement*, pp. 27–43. Elsevier, Amsterdam.

De Blust, G., Kuijken, E. and Paelinx, D. 1995. The Green Main Structure for Flanders. *Landschap*, **12**(3), 89–98.

De Lima Pufal, V., Buschbacher, R. and Garcia, A. 2000. *Brazil: Cerrado*. In A., Wood, P. Stedman-Edwards, and J. Mang, (eds.), *The Root Causes of Biodiversity Loss*, pp. 95–125. WWF, Earthscan, London and Sterling, VA.

De Wit, C. T. 1992. Resource use efficiency in agriculture. *Agricultural Systems*, **40**, 125–51.

Devillers, P. and Devillers-Terschuren, J. 1996. *A Classification of Palaearctic Habitats*. Nature and Environment No. 78. Council of Europe Publishing, Strasbourg.

DeYoung, R. and Kaplan, S. 1988. On averting the tragedy of the commons. *Environmental Management*, **12**, 273–83.

Di Ludovico, D. and Romano, B. 2000. The evaluation of environmental fragmentation using GIS techniques. *Planeco Newsletter*, **5**, 7–8.

Dobson, A., Ralls, K., Foster, M., Soulé, M., Simberloff, D., Doak, D. *et al.* 1998. Corridors: reconnecting fragmented landscapes. In M. E. Soulé and J. Terborgh (eds.), *Continental Conservation: Scientific Foundations of Regional Reserve Networks*, The Wildlands Project and Island Press, Washington, DC.

Dodd, C. K. Jr 1990. Effects of habitat fragmentation on a stream-dwelling species, the flattened musk turtle *Sternotherus depressus*. *Biological Conservation*, **54**, 33–45.

Dominguez, D. and Mariotti, D. 2001. Alianzas estratégicas: Nuevas dinámicas de acción en el conflicto por la tierra. Universidad Nacional de Buenos Aires, Facultad de Ciencias Sociales, Instituto Gino Germani. *Wayruro*, Fundación Wayruro, Jujuy, Sept.–Oct. 2001. (http:// www.wayruro.com.ar/wayru-c.html)

Eason, T. H. 2000. *Black bear status report, October 12, 2000*. Florida Fish and Wildlife Conservation Commission, Division of Wildlife. Tallahassee, FL.

Eason, T. H. and O'Meara, T. E. 2000. Comprehensive statewide bear management program in Florida. In *Managing Wildlife in the Face of Urbanisation*. The Florida Chapter of the Wildlife Society April 10–12, 2000 Symposium, Daytona Beach, Florida, p. 18.

Eisenberg, J. F. 1986. *The biological basis for corridors*. ENFO (Environmental Information Center of the Florida Conservation Foundation, Inc), Winter Park, FL.

Elbersen, B. and Prados, M. J. 1999. Desarrollo rural y calidad de vida en el entorno del Parque Nacional de Doñana. *Revista de Estudios Regionales*, **55**, 47–76.

Erickson, D. and Louisse, A. 1997. *Greenway Implementation in Metropolitan Regions: A Comparative Case Study of North American Examples*. University of Michigan, School of Natural Resources and Environment.

Estonian Ministry of the Environment 1997. *Estonian Environmental Strategy*. Estonian Ministry of the Environment, Tallin. (http://www.envir.ee/)

  1998. *Estonian Environmental Action Plan*. Estonian Ministry of the Environment, Tallin.

  2000. *Estonia – Vision 2010*. Estonian Ministry of the Environment, Tallinn.

Eupen, M. van and van der Veen, E. 1995. *Versnippering van ecologische infrastructuur in relatie tot verkeersafwikkeling in landelijke gebieden; een gebiedsgerichte benadering in de gemeente Opsterland*. Department of Physical Planning and Rural Development, Agricultural University, Wageningen.

European Commission. 1999. *ESDP (European spatial development perspective). Towards balanced and sustainable development of the territory of the European Union: agreed at the informal council of ministers responsible for spatial planning in Potsdam*, May 1999. Office for Official Publications of the European Communities, Luxemburg.

Evink, G. L. 1990. Wildlife crossings of Florida I-75. *Transportation Research Record*, **1279**, 54–9.

Evink, G., Garrett P. and Zeigler, D. eds. 1998. *Proceedings of the Second International Conference on Wildlife Ecology and Transportation*, Feb. 10–12, 1998 in Ft. Myers, FL. Florida Department of Transportation, Tallahassee, FL.

  eds. 1999. *Proceedings of the Third International Conference on Wildlife Ecology and Transportation*, Sept. 13–16, 1999 in Missoula, MT. Florida Department of Transportation, Tallahassee, FL.

Evink, G. L., Garrett, P., Zeigler, D. and Berry, J. eds. 1996. *Trends in addressing transportation related wildlife mortality: Proceedings of the transportation related wildlife mortality seminar*, June-1996. Florida Department of Transportation, Tallahassee, FL.

Fabos, J. 1995. Introduction and Overview: The Greenway Movement, Uses and Potentials of Greenways. In J. Fabos and J. Ahern (eds.), *Greenways: the Beginning of an International Movement, Landscape and Urban Planning*, **33**, 1–13.

Fabos, J. and Ahern, J. eds. 1995. *Greenways: the Beginning of an International Movement. Landscape and Urban Planning*, **33**, 1–491.

Fabos, J., Lindhult, M. and Ryan, R. 1999. *The New England Greenway Vision Plan*. (http://www.umass.edu/greenway)

Fabos, J. G. 1985. *Land-Use Planning: From Global to Local Challenge*. Chapman and Hall, New York.

Farhig, L. and Merriam, G. 1985. Habitat patch connectivity and population survival. *Ecology*, **66**, 1762–8.

Fahrig, L., Pedlar, J., Pope, S., Taylor, P. and Wegner, J. 1995. Effect of road traffic on amphibian density. *Biological Conservation*, **73**, 177–82.

FDOT (Florida Department of Transportation) 1994. *Connections, bringing Florida together: 2020 Florida transportation plan initial draft report*. Florida Department of Transportation, Tallahassee, FL.

  1995. *The 2020 Florida Transportation Plan, Connections: bringing Florida together*. Florida Department of Transportation, Tallahassee, FL.

  1998. *Historical Report on Mileage and Traffic on Florida's State Highway System*. Florida Department of Transportation, Tallahassee, FL.

Ferrer, M., Prat, N., Toja, J., Baltanás, A., Merino, J., Murillo, J. M. *et al*. 2002. *Seguimiento de la contaminación por metales pesados en seres vivos de Doñana y su entorno: efectos subletales a nivel de individuo e impacto en las poblaciones. Informe Final*. Estación Biológica de Doñana, Sevilla.

FFWCC (Florida Fish and Wildlife Conservation Commission) 2000. Unpublished data. Florida Fish and Wildlife Conservation Commission, Tallahassee, FL.

FGP (Florida Greenways Project) 1999. *Executive summary of the Florida statewide greenways system planning project: recommendations for the physical design of a statewide greenways system.* Florida Greenways Project, Department of Landscape Architecture, University of Florida, Gainesville, FL.

FHWA (Federal Highway and Waterway Administration) 1996. *Highway Statistics Summary – to 1995.* Office of Highway Information Management, Federal Highway & Waterway Administration, Washington, DC. (http://www.bts.gov/site/news/fhwa/HighwayStats-Summary95/)

Flink, C. A. and Searns, R. 1993. *Greenways: A Guide to Planning, Design, and Development.* Island Press, Washington, DC.

Florida Audubon Society. 2001. Rural and Family Lands Protection Act Passes Legislature. *Audubon Advocate,* Issue 9. Winter Park, FL.

Florida Department of Environmental Protection 1998. *Ecosystem management at work in Florida.* Florida Department of Environmental Protection, Tallahassee, FL.

Florida Forever Advisory Council. 2001. *Final Report.* Florida Department of Environmental Protection, Tallahassee, FL.

Florida Greenways Commission 1994. *Creating a statewide Greenways system, for people...for wildlife...for Florida.* Report to the Governor. Florida Department of Environmental Protection, Tallahassee, FL.

Florida Natural Areas Inventory 1997. *Plants and lichens, vertebrates, and natural communities tracked by Florida Natural Areas Inventory.* Florida Natural Areas Inventory, Tallahassee, FL.

Florida Resources and Environmental Information Center. 1990. *The Fifty Major Rivers of Florida.* Florida State University, Tallahassee, FL.

Fluit, N. van der, Cuperus, R. and Canters, K. J. 1990. *Mitigerende en compenserende maatregelen aan het hoofdwegennet voor het bevorderen van natuurwaarden.* CML mededelingen 65 Centrum voor Milieukunde, Rijksuniversiteit, Leiden.

Ford, S. G. 1980. *Evaluation of highway deer kill mitigation on SIE/LAS-395.* Final report no. FHWA/CA/TP-80-01. California Department of Transportation. Sacramento, CA.

Forman, R. T. T. 1983. Corridors in a landscape: their ecological structure and function. *Ekologia (Czechoslovakia),* **2,** 375–87.

1987. The ethics of isolation, the spread of disturbance, and landscape ecology. In M. G. Turner (ed.), *Landscape Heterogeneity and Disturbance,* pp. 213–29. Springer-Verlag, New York.

1995. *Land Mosaics: the Ecology of Landscapes and Regions.* Cambridge University Press, Cambridge, UK.

Forman, R. T. T. and Alexander, L. E. 1998. Roads and their major ecological effects. *Annual Review of Ecology and Systematics,* **29,** 207–31.

Forman, R. T. T. and Godron, M. 1986. *Landscape Ecology.* J. Wiley & Sons, New York.

Forman, R. T. T. and Hersperger, A. M. 1997. Ecologia del paesaggio e pianificazione, una potente combinazione. *Urbanistica,* **108,** 61–6.

Foruseth, O. J. and Altman, R. E. 1991. Who's on the greenway: socioeconomic, demographic, and locational characteristics of greenway users. *Environmental Management,* **15,** 329–36.

Foster, M. L. and Humphrey, S. R. 1995. Use of highway underpasses by Florida panthers and other wildlife. *Wildlife Society Bulletin,* **23,** 95–100.

Friedman, D. S. 1997. *Nature as infrastructure: The National Ecological Network and wildlife crossing structures in the Netherlands.* Report 138. DLO Winand Staring Centre, Wageningen.

Fry, G. and Gustavsson, R. 1996. Testing landscape design principles: the Landscape Laboratory. In R. H. G. Jongman (ed.), *Ecological and Landscape Consequences of Land Use Change,* pp.143–54. ECNC, Tilburg.

Gallart, F., Benito, G., Martín-Vide, J. P., Benito, A., Prió, J. M. and Regüés, D. 1999. Fluvial geomorphology in the dispersal and fate of pyrite mud particles released by the

Aznalcóllar mine tailings spill. *The Science of the Total Environment*, **242**, 13–26.

Gavrin, B. J., Rasmussen, M., Fabos, J. G. and Ahern, J. 1993. *A Management Plan to Balance Cultural and Natural Resources: The Minute Man National Historic Park Case Study*. Research Bulletin 744, Massachusetts Agricultural Experiment Station, University of Massachusetts, Amherst, MA.

Gilbert, T. 1996. An overview of black bear roadkill in Florida 1976–1995. In G. Evink, P. Garrett, D. Zeigler, and J. Berry (eds.), *Trends in addressing transportation related wildlife mortality: Proceedings of the transportation related wildlife mortality seminar*, June 1996. Florida Department of Transportation, Tallahassee, FL.

Gilbert, T. and Wooding, J. 1994. *Chronic roadkill problem areas for black bear in Florida; A preliminary report of transportation impacts on wildlife*. Florida Game and Freshwater Fish Commission, Tallahassee, FL.

Gill, F. B. 1995. *Ornithology*. Second Edition. W. H. Freeman and Company, New York.

Gobster, P. H. 1995. Perception and use of a metropolitan greenway system for recreation. *Landscape and Urban Planning*, **33**, 401–13.

Goggin, M. L., Bowman, A. O., Lester, J. P. and O'Toole, L. J. Jr. 1990. *Implementation Theory and Practice: Toward a Third Generation*. Harper Collins, Inc.

Green, M., Alerstam, T., Clausen, P., Drent, R. and Ebbinge, B. S. 2000. Site use by dark – bellied brent geese *Branta bernicla bernicla* on the Russian tundra as recorded by satellite telemetry: implications for East Atlantic Flyway conservation. *Wildlife Biology*, **8**, 229–39.

2002. Dark-bellied brent geese *Branta bernicla bernicla*, as recorded by satellite telemetry, do not minimize flight distance during spring migration. *Ibis*, **144**, 106–21.

Greenways and Natural Areas Collaborative 1997. *Metro Greenprint*. Minnesota Department of Natural Resources, St. Paul, MN.

Grift, R. 2001. *How fish benefit from floodplain restoration along the lower River Rhine*. Ph.D. Thesis, Wageningen University.

Grimalt, J. O., Ferrer, M. and Macpherson, E. 1999. The mine tailing accident in Aznalcóllar. *The Science of the Total Environment*, **242**, 3–11.

Groot Bruinderink, G. van der Sluis, T., Lammertsma, D. and Kuipers, H. 2001. Edel boegbeeld voor het Ketelwald – Het edelhert als basis voor de nadere uitwerking van het Ketelwald als onderdeel van de Noordwesteuropese Ecologische Hoofdstructuur. *Nieuwe Wildernis*, **6**(21–22), 32–7.

Groot Bruinderink, G., van der Sluis, T., Lammertsma, D., Opdam, P. and Pouwels, R. 2003. Designing a coherent ecological network for large mammals in Northwestern Europe. *Conservation Biology*, **17**(2), 549–57.

Grove, N. 1990. Greenways: Paths to the future. *National Geographic*, **7**, 77–98.

Grove, N. 1990. Greenways: Patns to the future. *National Geographic*, **190**(6), 77–98.

Gustafson, E. 1998. Quantifying landscape spatial pattern: what is the state of the art? *Ecosystems*, **1**, 143–56.

Haapoja, M. 1994. Conservation easements: are they for you? *American Forests*, **100**(1–2), 29–32.

Hanski, I. 1994. A practical model of metapopulation dynamics. *Journal of Animal Ecology*, **63**, 151–62.

1997. Metapopulation dynamics – From concepts and observations to predictive models. In I. Hanski and M. E. Gilpin, (eds.), *Metapopulation Biology*, pp. 69–91. Academic Press, San Diego.

Hanski, I. and Gilpin, M. 1991. Metapopulation dynamics: Brief history and conceptual domain. In M. Gilpin and I. Hanski (eds.), *Metapopulation Dynamics: Empirical and Theoretical Investigations*, pp. 3–16. Academic Press, London.

(eds.) 1997. *Metapopulation Biology*. Academic Press, San Diego.

Hanski, I., Poyry, J., Pakkala, T. and Kuussaari, M. 1985. Multiple equilibria in metapopulation dynamics. *Nature*, **377**, 616–21.

Harris, L. D. 1984. *The Fragmented Forest. Island Biogeography Theory and the Protection of Biotic Diversity*. University of Chicago Press, Chicago, IL.

1985. *Conservation corridors: a highway system for wildlife*. ENFO Report 85–5, Florida Conservation Foundation, Winter Park, FL.

Harris, L. D. and Atkins, K. 1991. Faunal movement corridors in Florida. In W. E. Hudson (ed.), *Landscape Linkages and Biodiversity*, pp. 117–34. Island Press, Defenders of Wildlife, Washington, DC.

Harris, L. D. and Gallagher, P. B. 1989. New initiatives for wildlife conservation: The need for movement corridors. In G. Mackintosh (ed.), *Defense of Wildlife: Preserving Communities and Corridors*, pp. 11–24. Defenders of Wildlife, Washington, DC.

Harris, L. D., Hoctor, T. S. and Gergel, S. E. 1996a. Landscape processes and their significance to biodiversity conservation. In O. E. Rhodes Jr., K. Chesser and M. H. Smith (eds.), *Population Dynamics in Ecological Space and Time*, pp. 319–47. The University of Chicago Press, Chicago, IL.

Harris, L. D., Hoctor, T., Maehr, D. and Sanderson, J. 1996b. The role of networks and corridors in enhancing the value and protection of parks and equivalent areas. In R. G. Wright (ed.), *National Parks and Protected Areas: their Role in Environmental Areas*, pp. 173–198. Blackwell Science, Cambridge, MA.

Harris, L. D. and Scheck, J. 1991. From implications to applications: the dispersal corridor approach to the conservation of biological diversity. In D. A. Saunders and R. J. Hobbs (ed.), *Nature Conservation 2: the Role of Corridors*, pp. 189–220. Surrey Beatty and Sons, Chipping Norton, New South Wales, Australia.

Harris, L. D. and Silva-Lopez, G. 1992. Forest fragmentation and the conservation of biological diversity. In P. Fielder and S. Jain (eds.), *Conservation Biology: the Theory and Practice of Nature Conservation*, pp. 197–237. Chapman and Hall, New York.

Harrison, J. 1998. *Information Strategy for the Pan-European Ecological Network (Project STRA-REP 10). Project Concept*. Council of Europe, Committee of Experts for the European Ecological Network. STRA-REP (98) 11.

Harrison, S. and Bruna, E. 1999. Habitat fragmentation and large-scale conservation: what do we know for sure? *Ecography*, **22**, 225–32.

Hellgren, E. C. and Maehr, D. S. 1992. Habitat fragmentation and black bears in the eastern United States. In E. P. Orff (ed.), *Proceedings of the 11th Eastern Black Bear Workshop*, Waterville Valley, NH, pp. 154–165. New Hampshire Dept. of Fish and Game, Concord, NH.

Hellmund Associates 1998. *Planning Trails with Wildlife in Mind: A Handbook for Trail Planners*. Trails and Wildlife Task Force, Colorado State Parks.

Hess, G. 1994. Conservation corridors and contagious disease: a cautionary note. *Conservation Biology*, **8**, 256–62.

Hewitt, D., Cain, A., Tuovilla, V., Shindle, D. and Tewes, M. 1998. Impacts of an expanded highway on ocelots and bobcats in southern Texas and their preferences for highway crossings. In G. Evink, P. Garrett, D. Zeigler and J. Berry (eds.), *Proceedings of the International Conference on Wildlife Ecology and Transportation*, Feb. 10–12, pp. 126–34. Florida Department of Transportation, Tallahassee, FL.

Hobbs, R. J. 1992. The role of corridors in conservation: solution or bandwagon? *Trends in Ecology and Evolution*, **7**, 389–92.

Hoctor, T. S., Carr, M. H. and Zwick, P. D. 2000. Identifying a linked reserve system using a regional landscape approach: the Florida ecological network. *Conservation Biology*, **14**(4), 984–1000.

Hoover, A. P. and Shannon, M. A. 1995. Building greenway policies within a participatory democracy framework. *Landscape and Urban Planning*, **33**, 433–59.

Houweling, H., Jansen, M. J. W., Kalkhoven, J. T. R. and Pouwels, R. 1999. *Gevoeligheidsanalyse op basis van de studie DELTA-ECONET*. Internal report, Alterra Wageningen.

Howard, E. 1898. *Tomorrow: A Peaceful Path to Reform*. Swan Sonnenschein, London. (Second edition, 1945, *Garden Cities of Tomorrow*, Faber and Faber, London.)

Hubbell, S. P. 2002. *Unified Neutral Theory of Biodiversity and Biogeography*. Princeton University Press, Princeton, NJ.

Hubert, B. 1991. Changing land uses in the Provence (France); Multiple land use as a management tool. In J. Baudry and R. G. H. Bunce (eds.), *Land Abandonment and its Role in Conservation* (Options Mediterranéennes Série Séminaires 15), pp. 31–52. CIHEAM, Zaragoza.

Hunt, A., Dickens, H. J. and Whelan, R. J. 1987. Movement of mammals through tunnels under railway lines. *Australian Zoology*, **24**, 89–93.

INDEC 2000a. *Situación y evolución social provincial. Jujuy. Síntesis No. 1*. Ministerio de Economía, República Argentina, Buenos Aires.

2000b. *Situación y evolución social provincial. Salta. Síntesis No. 1*. Ministerio de Economía, República Argentina, Buenos Aires.

Interpretation Manual of European Union Habitats 1999. *EUR* 15/2, European Commission DG Environment, Brussels.

IUCN. 1992. *The Rio Declaration on the Environment*. (http://www.iucn.org)

IUCN-CNPPA, 1982: *United Nations List of National Parks and Protected Areas*. IUCN, Gland.

Ivy, R. 1992. Front porch on the river. *Landscape Architecture*, **82**(12), 70–1.

Jaarsma, C. F. 1997. Approaches for the planning of rural road networks according to sustainable land use planning. *Landscape and Urban Planning*, **39**, 47–54.

2000. *Sustainable Land Use Planning and Planning of Rural Road Networks*. Agricultural Engineering International: the CIGR Electronic Journal of Scientific Research and Development, volume II, December. (http://www.agen.tamu.edu.cigr/)

Jaarsma, C. F. and Baltjes, C. R. 1995. *Herinrichting Kust West Zeeuwsch-Vlaanderen: ontsluiting onder de loep. Eindrapport: varianten voor een ontsluitingsstructuur met verkeersprognose en effectbepaling*. Nota vakgroep Ruimtelijke Planvorming 59, Wageningen.

Jaarsma, C. F., Luimstra, J. O. K. and de Wit, T. J. 1995. *De kortste weg naar een verkeersleefbaar platteland. Onderzoek ruraal verblijfsgebied Ooststellingwerf*. Nota

vakgroep Ruimtelijke Planvorming 58, Wageningen.

Jaarsma, C. F. and van Langevelde, F. 1996. The motor vehicle and the environment: balancing between accessibility and habitat fragmentation. In *29th ISATA, dedicated Conference on the Motor Vehicle and the Environment*, Florence, pp. 299–306. Automotive Automation Ltd, Croydon.

1997. *Right-of-way Management and habitat fragmentation: an integral approach with the spatial concept of the Traffic Calmed Rural Area*. 6th International Symposium Environmental Concerns in Rights-of-way Management, New Orleans (LA), 24–26 February, pp. 383–92. Elsevier, Amsterdam.

Jackson, S. 1996. Underpass systems for amphibians. In G. Evink, P. Garrett, D. Zeigler and J. Berry (eds.), *Trends in Addressing Transportation Related Wildlife Mortality*, Florida Department of Transportation, Tallahassee, FL.

Jacobson, L. 1997. Tennessee triumph. *Planning*, **63**(5), 20–2.

Jagomägi, J. 1983. Ecologically balanced land. *Eesti Loodus*, **26**(4), 219–24.

Jagomägi, J. Mardiste, H., Peterson, U., Roosaare, J. and Sepp, K. 1985. *Methods and Practice of Territorial Planning for Nature Protection*. Final report of contract A-80/9. Tartu.

Jagomägi, J., Oja, T., Sepp, K., Sillaots, T., Tambets, J. and Tammemäe, O. 2000. Green network. In *Estonia – Vision 2010*, pp. 57–63. Estonian Ministry of the Environment, Tallinn.

Jehl, Do. January 18, 2001. Norten Hearings to Pose Test for Bush's Plan to Change Policies on Use of Public Lands. *New York Times*.

Jones, F. R. H. 1968. *Fish Migration*. The Camelot Press Ltd., London and Southampton.

Jongman, R. H. G. 1992. Vegetation, river management and land use in the Dutch Rhine floodplains. *Regulated Rivers*, **7**, 279–89.

1995. Nature conservation planning in Europe: developing ecological networks. *Landscape and Urban Planning*, **32**, 170–83.

1998a. Le reti ecologiche, a quale scopo? *Naturopa*, **87**, 4–5.

1998b. The role of rivers in the European Ecological Network. In: P. H. Nienhuis and R. S. E. W. Leuven (eds.), *New Concepts for River Systems*, pp. 53–66. Backhuys Publishers, Leiden.

1999. Application of ecological principles: Ecological Networks. In J. A. Wiens and M. R. Moss (eds.), *Landscape Ecology: the Science and the Action. Book of abstracts*, pp. 77–8. International Association for Landscape Ecology, Guelph.

Jongman, R. H. G. and Kristiansen, I. 2001. *National and Regional Approaches for Ecological Networks in Europe.* Council of Europe Publishing, Nature and Environment No. 110. Strasbourg.

Jongman, R. H. G. and Leemans, J. A. A. M. 1982. *Vegetatieonderzoek Gelderse uiterwaarden*. Provincie Gelderland.

Jongman, R. H. G., Ter Braak, C. J. F. and Van Tongeren, O. F. R., 1995. *Data Analysis in Community and Landscape Ecology.* Cambridge University Press, Cambridge, UK.

Jongman, R. H. G. and Troumbis, A. Y. 1995. *The Wider Landscape for Nature Conservation: Ecological Corridors and Buffer Zones.* MN2.7 Project-report 1995, submitted to the European Topic Centre for Nature Conservation in Fulfilment of the 1995 Work Programme. ECNC, Tilburg.

Jungwirth, M., Schmutz, S. and Weiss, S. (eds.) 1998. *River Continuum and Fish Migration – Going Beyond the Longitudinal River Corridor in Understanding Ecological Integrity.* Marston Book Services, Ltd.

Junk, W. J., Bayley, P. B. and Sparks, R. E. 1989. The flood pulse concept in river-floodplain systems. In Dodge, D. P. (ed.), *Canadian Special Publication of Fisheries and Aquatic Sciences*, **106**, 110–27.

Jurajda, P. 1995. *Effect of channelization and regulation on fish recruitment in a flood plain river.* Regulated Rivers: Research and Management, Vol. 10: 207–15.

Kaplan, R., Kaplan, S. and Ryan, R. L. 1998. *With People in Mind: Design and Management of Everyday Nature.* Island Press, Washington, DC.

Kautz, R. 1992. Florida wildlife habitat: The last 50 years. *Florida Wildlife*, September–October, pp. 2–6.

1993. Trends in Florida wildlife habitat 1936–1987. *Florida Scientist*, **56**, 7–24.

Kavaliauskas, P. 1995. The Nature Frame. *Landschap* **12**(3), 17–26.

1996. Lithuania, the nature frame. In P. Nowicki *et al.* (eds.) *Perspectives on Ecological Networks*, pp. 93–101. ECNC, Tilburg.

Klingebiel, A. A. and Montgomery, P. H. 1961. *Land-capability classification. Agricultural Handbook 210, Soil Conservation Service.* US Government Printing Office, Washington, DC.

Koolstra, B. and Reijnen, R. 1998. *Evaluatie van de ecologische verbindingszones in de provincie Gelderland.* Rapport 372. IBN-DLO, Wageningen.

Kubes, J. 1996. Biocentres and corridors in a cultural landscape: a critical assessment of the territorial system of ecological stability. *Landscape and Urban Planning*, **35**, 231–40.

Külvik, M. 1998. Synergies are taking us further. *Naturopa (Special Edition: Ecological Networks)*, **87**, 28–9.

Labaree, J. M. 1992. *How Greenways Work: A Handbook on Ecology.* Second edition. National Park Service and Atlantic Center for the Environment, Ipswich, MA.

Lambeck, R. J. 1997. Focal species: a multi-species umbrella for nature conservation. *Conservation Biology*, **11**, 849–56.

Lambright, W. H. 1997. The rise and fall of interagency cooperation: the U.S. global climate change research program. *Public Administration Review*, **57**(1), 36–44.

Land Trust Alliance. 1999. *Voters Invest in Parks and Open Space.* 1998 Referenda Results.

2000. *Voters Invest in Parks and Open Space*, 1999 Referenda Results. (www.lta.org)

2001. *Voters Invest in Open Space*, 2000 Referenda Results.

Land, D. and Lotz, M. 1996. Wildlife crossing designs and use by Florida panthers and other wildlife in southwest Florida. In G. Evink, P. Garrett, D. Zeigler and J. Berry, (eds.), *Trends in Addressing Transportation Related Wildlife Mortality.* Florida Department of Transportation, Tallahassee, FL.

Lande, R. 1988. Demographic models of the northern spotted owl (*Strix occidentalis caurina*). *Oecologia* **75**, 601–7.

Langevelde, F. van 1994. Conceptual integration of landscape planning and landscape ecology, with a focus on the Netherlands. In E. A. Cook and H. N. van Lier (eds.), *Landscape Planning and Ecological Networks. Developments in Landscape Management and Urban Planning*, pp. 27–70. Elsevier, Amsterdam.

1999. *Habitat connectivity and fragmented nuthatch populations in agricultural landscapes*. Ph. D. Thesis, Wageningen Agricultural University.

Langevelde, F. van and Jaarsma, C. F. 1997. Habitat fragmentation and infrastructure: the role of minor rural roads and their traversability. In K. Canters, (ed), *Habitat Fragmentation and Infrastructure. Congress on Habitat Fragmentation and Infrastructure, Maastricht, The Hague, The Netherlands, 1995*, pp. 171–82. Ministry of Transport, Public Works and Water Management, Delft.

Langevelde, F. van, Schotman, A., Claassen, F. and Sparenburg, G. 2000. Competing land use in the reserve site selection problem. *Landscape Ecology*, **15**(3), 243–56.

Langton, T. E. S. (ed.) 1989. *Amphibians and Roads*. ACO Polymer Products Ltd, Bedfordshire.

Lankester, K., van Apeldoorn, R., Meelis, E. and Verboom, J. 1991. Management perspectives for populations of the Eurasian badger (*Meles meles*) in a fragmented landscape. *Journal of Applied Ecology*, **28**, 561–73.

Leedy, D. L. 1975. *Highway-wildlife relationships*. Vol. 1. *A state of the art report*. FHWA-RD-76-4.

Lerner, S. and Poole, W. 1999. *The Economic Benefits of Parks and Open Space*. The Trust for Public Land, San Francisco.

Levins, R. 1970. Extinction. In M. Gerstenhaber (ed.), *Some Mathematical Problems in Biology*, pp. 77–107. American Mathematical Society, Providence, RI.

Levy, G., Ruggeri, A., Santucho, M., de Gori, E. and Lopez, E. 1999. *La reordenación socio-económica y espacial de las familias de Alto Calilegua: ¿desestructuración o modalidad estratégica de acceso y control de los recursos?* Report presented to the 2nd International Workshop of the Bioandes Project, Calilegua city, Jujuy.

Lewis, P. H. Jr. 1964. Quality Corridors for Wisconsin. *Landscape Architecture Quarterly*, Louisville, Kentucky, January, pp. 101–8.

1996. *Tomorrow by Design: A Regional Design Process for Sustainability*. John Wiley & Sons, Inc., New York.

Lier, H. N. van 1995. Natural resource management in land use planning. In *Tessa program Protection and sustainable development of rural areas*, lecture 3. Wageningen Agricultural University, Department of Physical Planning.

Lilleleht, V. (ed.) 1998. *Red Data Book of Estonia*. Tartu.

Limpens, H. J. G. A. and Kapteijn, K. 1991. Bats, their behaviour and linear landscape elements. *Myotis*, **29**, 63–71.

Lindblom, C. E. 1959. The science of 'muddling through.' *Public Administration Review*, **39**, 517–26.

Lindenmayer, D. B. and Possingham, H. P. 1994. *The risk of extinction: ranking management options for Leadbeater's possum using PVA*. Centre for Resource and Environmental Studies, ANU, Canberra, Australia.

1995. Modelling the viability of metapopulations of the endangered Leadbeater's possum in south-eastern Australia. *Biodiversity and Conservation*, **4**, 984–1018.

Lindsey, G. 1999. Use of urban greenways: insights from Indianapolis. *Landscape and Urban Planning*, **45**, 145–57.

Lindsey, G. and Knaap, G. 1999. Willingness to pay for urban greenway projects. *Journal of the American Planning Association*, **65**, 297–313.

Linehan, J., Gross, M. and Finn, J. 1995. Greenway planning: developing a landscape ecological network approach. *Landscape and Urban Planning*, **33**, 179–93.

Lipsky, Z. 1992. Use of historic documents about territory for study of landscape development. In: *Proceedings of the field workshop Ecological Stability of Landscape Ecological Infrastructure Ecological Management*, pp. 80–87. Federal Committee for the Environment, Institute of Applied Ecology, Kostelec nad Cerny lesy.

Little, C. E. 1990. *Greenways for America*. Johns Hopkins University Press, Baltimore.

LoBuono, J. P. 1988. Alligator Alley: Protecting natural habitat. *Journal of the Florida Engineering Society*, February 1988, 14–16.

Logan, T. and Evink, G. 1985. Safer travel for the Florida panther. *Florida Naturalist*, Spring 1985, 6–7.

Lowry, W. R. 1998. *Preserving Public Lands for the Future: The Politics of Intergenerational Goods*. Georgetown University Press, Washington, DC.

Lucas, P. and van Oort, G. 1993. *Dynamiek in een stadsrandzone-Werken en wonen in de stadsrandzone van de agglomeratie Utrecht*. Utrecht, Faculteit Ruimtelijke Wetenschappen, Rijksuniversiteit Utrecht.

Lucretius, 1937 (iii ed.). *De Rerum Natura*. Transl. by W. H. D. Rouse. The Loeb Classical Library. William Heinemann, London.

Lynch, K. 1976. *Managing the Sense of a Region*. The MIT Press, Cambridge, MA.

Mabelis, A. 1990. Natuurwaarden in cultuurlandschappen. *Landschap*, 7, 253–67.

MacArthur, R. H. and Wilson, E. O. 1967. *The Theory of Island Biogeography*. Princeton University Press, Princeton, NJ.

Machado, J. R., Andresen, M. T., Rico, A. T., Ahern, J. and Fabos, J. G. 1995. Metropolitan Landscape Planning: A Greenway Vision for the Metropolitan Area of Lisbon. *Landschap*, 95(3), 111–21.

MacKaye, B. 1928. *The New Exploration* (reprinted in 1962). University of Illinois Press, Urbana, IL.

Macpherson, G. 1993. *Highway and Transportation Engineering and Planning*. Longman, Harlow.

Mader, H. J. 1984. Animal habitat isolation by roads and agricultural fields. *Biological Conservation*, 29, 81–96.

1988. The significance of paved agricultural roads as barriers to ground dwelling arthropods. In K.-F. Schreiber (ed.), *Connectivity in Landscape Ecology. Proceedings of the 2nd International Seminar of the International Association for Landscape Ecology*, pp. 97–100. *Münsterische Geographische Arbeiten* 29. Münster.

Madsen, A. B. 1996. Otter *Lutra lutra* mortality in relation to traffic, and experience with newly established fauna passages at existing road bridges. *Lutra*, 39(2), 76–88.

Maehr, D. S. 1997a. *The Florida Panther: Life and Death of a Vanishing Carnivore*. Island Press, Washington, DC.

1997b. The comparative ecology of bobcat, black bear, and Florida panther in south Florida. *Bulletin of the Florida Museum of Natural History*, 40(1), 1–176.

Maehr, D. S. and Cox, J. A. 1995. Landscape features and panthers in Florida. *Conservation Biology*, 9, 1008–19.

Maehr, D. S., Land, E. D. and Roelke, M. E. 1991. Mortality patterns of panthers in southwest Florida. *Proceedings of the Annual Conference of the Southeastern Association of State Fish and Wildlife Agencies*, 45, 201–7.

Makhzoumi, J. and Pungetti, G. 1999. *Ecological Landscape Design and Planning: The Mediterranean Context*. Spon, London.

Mander, Ü. 1980. Designing of buffer zones for protected mires. In *Agriculture and Environment Protection*. Proceedings of the Regional Conference, Kingissepa, May 30–31, 1980, pp. 81–5. Academy of Science Publishing, Tallinn.

1983. *Some methods for ecological optimization of agricultural landscapes*. Ph.D. Thesis. Tartu State University. Tartu.

1995. Riparian buffer zones and buffer strips on stream banks: Dimensioning and efficiency assessment from catchments in Estonia. In M. Eiseltová and J. Biggs (eds.), *Restoration of Stream Ecosystems* (IWRB Publication No 37), pp. 45–64. Slimbridge, Gloucester.

Mander, Ü., Jagomägi, J. and Külvik, M. 1988. Network of compensative areas as an ecological infrastructure of territories. In K.-F. Schreiber (ed.), *Connectivity in Landscape Ecology. Proceedings of the 2nd International Seminar of the International Association for Landscape Ecology*, pp. 35–8 *Münsterische Geographische Arbeiten* 29. Münster.

Mander, Ü., Palang, H. and Jagomägi, J. 1995. Ecological networks in Estonia: impact of landscape change. *Landschap* (Special Issue) 12(3), 27–38.

Mann, C. C. 1995. Filling in Florida's gaps: species protection done right? *Science*, **269**: 318–20.

Manzanal, M. and Arrieta, J. 2000. *Diagnóstico socioeconómico del sector argentino de la cuenca del Río Bermejo*. In Report Final del Programa Estratégico de Acción para la Cuenca del Río Bermejo. Organizacion de Estados Americanos (OEA), Programa GEF PEA-Bermejo, Buenos Aires.

Manzano, M., Custodio, E., Ayora, C. and Navarrete, P. 2000. Contamination of the Guadiamar river aquifer after the Aznalcóllar mine accident, SW Spain. *Boletín Geológico y Minero*, **111**(6), 93–106.

Margoluis, R. and Margoluis, C. 2000. *What does it take to make conservation work? Conditions for success in conservation*. Lessons from the Field Series. Biodiversity Support Program, Washington, DC.

Marini, M. A., Robinson, S. K. and Heske, E. J. 1995. Edge effects on nest predation in the Shawnee-National-Forest, southern Illinois. *Biological Conservation*, **74**(3), 203–13.

Matus, R. 1999. Farms consumed by urban money. *The Gainesville Sun*. Gainesville, FL.

Mazmanian, D. A. and Sabatier, P. A. 1989. *Implementation and Public Policy*. University Press of America, Lanham.

McDowall, R. M. 1988. *Diadromy in Fishes – Migrations between Freshwater and Marine Environments*. Cambridge University Press, Cambridge, UK.

McHarg, I. I. 1969. *Design with Nature*. Natural History Press, Garden City, NY.

Meffe, G. K. and Carroll, C. R. 1994. *Principles of Conservation Biology*, pp. 254–6. Sinauer Associates, Sunderland. MA.

Meffe, G. K., Carroll, C. R. *et al.* 1997. *Principles of Conservation Biology*, Second edition. Sinauer Associates, Sunderland, MA.

Meiner, A. (ed.) 1999. *Land Cover of Estonia*. Estonian Environment Information Center, Tallinn.

Metro 1992. *Metropolitan Greenspaces Master Plan Summary*. Metro, Portland, OR.

1999. *Metro's open spaces and land acquisition: Report to citizens*. Metro, Portland, OR.

Metropolitan Council 1991. *Recreation open space development guide/policy plan*. Metropolitan Council. Publication No. 580-91-051, St. Paul, MN.

1995. *Regional recreation open space capital improvement program 1996–2005*. Metropolitan Council Publication No. 78-95-022, St. Paul, MN.

Miklós, L. 1996. The concept of the territorial system of landscape stability in Slovakia. In R. H. G. Jongman (ed.), *Ecological and Landscape Consequences of Land Use Change in Europe*, vol. 2, pp. 385–406. ECNC, Tilburg.

Milos, M., Hsieh, J. and Erickson, D. 1995. Greenway infrastructure planning from a local perspective. In J. L. Thompson, D. W. Lime, B. Gartner and W. M. Sames (eds.), *Proceedings of the Fourth International Outdoor Recreation and Tourism Trends Symposium and the 1995 National Recreation Resource Planning Conference*, pp. 561–6. University of Minnesota, St Paul, MN.

Ministerie Landbouw, Natuurbeheer en Visserij 1997. *Programma Beheer. Het beheer van natuur, bos en landschap binnen en buiten de Ecologische Hoofdstructuur*. Den Haag.

2000. *Natuur voor mensen mensen voor natuur. Nota natuur bos en landschap in de 21e eeuw*. Den Haag.

Ministero dell'Agricoltura e Foreste 1976. *Carta della Montagna*. Geotecneco, Pesaro.

Ministry of Agriculture 1989. Ley 4/89 de 27 de marzo (BOE no 74 de 28-3-89).

Ministry of Transport 1996. *Defragmentation*. Ministry of Transport of The Netherlands, The Hague.

Möller, K. (comp.) 1998. *CORINE Biotopes Estonia*. Tallinn.

Moore, R. L. and Graefe, A. R. 1994. Attachments to recreation settings; The case of rail-trail users. *Leisure Sciences*, **16**, 17–31.

Moore, R. L., Graffe, A. R., Gitelson, R. J. and Porter, E. 1992. *The Impacts of Rail-Trails: A Study of the Users and Property Owners From Three Trails*. National Park Service, Rivers, Trails, and Conservation Assistance Program, Washington, DC.

Morisi, A. (ed.) 2001. *Recupero e Gestione Ambientale della Pianura. La Rete Ecologica del Persicetano*. Centro Agricoltura Ambiente, Crevalcore (Bologna).

Mücher, C. A. (ed.) 2000. *PELCOM Project*. Final Report, February 2000. Alterra Wageningen.

Muskett, C. J. and Jones, M. P. 1980. The dispersal of lead, cadmium, and nickel from motor vehicles and effects on roadside invertebrate macrofauna. *Environmental Pollution*, **23**, 231–42.

Mwamfupe, D. 1998. *Demographic impacts on protected areas in Tanzania and options for action*. PARKS, Vol. 8, No. 1, February 1998. Protected Area Programme, IUCN, Gland.

Myers, R. L. 1990. Scrub and high pine. In R. L. Myers and J. J. Ewel, (eds.), *Ecosystems of Florida*, pp. 150–93. The University of Central Florida Press, Orlando, FL.

Nadir, A. and Chafatinos, T. 1990. *Los suelos del NOA (Salta y Jujuy)*, 3 vols., Salta, Argentina.

Naiman, R. J., Melillo, J. M., Lock, M. A., Ford, T. E. and Reice, S. R. 1987. Longitudinal patterns of ecosystem processes and community structure in a subarctic river continuum. *Ecology*, **68**, 1139–56.

National Park Service 1991. *Economic Impacts of Protecting Rivers, Trails, and Greenway Corridors: A Resource Book*. National Park Service. Rivers, Trails and Conservation Assistance Program, Washington, DC.

1992. *The Impacts of Rail-Trails: A Study of Users and Nearby Property Owners from Three Trails*. National Park Service. Rivers, Trails and Conservation Assistance Program, Washington, DC.

1995. *Greenways for the Southeast Tennessee River Valley*. National Park Service, Atlanta, GA.

Newton, N. T. 1971. *Design on the Land*. Harvard University Press, Cambridge, MA.

Nicolas, J. C. and Steiner, R. L. 2000. Growth Management and Smart Growth in Florida. *Wake Forest Law Review*, **35**(3), 645–70.

Northcote, T. G. 1978: *Migratory strategies and production in freshwater fishes*. In S. D. Gerking (ed.), *Ecology of Freshwater Fish Production*, John Wiley and Sons, New York.

Northeast Illinois Planning Commission and Openlands Project 1992. *Northeast Illinois Regional Greenways Plan*. Northeastern Illinois Planning Commission and Openlands Project, Chicago, IL.

1994. *State of the Greenways Report: A Report on Implementation of the Northeastern Illinois Greenways Plan*. Northeastern Illinois Planning Commission and Openlands Project, Chicago, IL.

1997. *Northeast Illinois Regional Greenways and Trails: Implementation Program*. Northeastern Illinois Planning Commission and Openlands Project, Chicago, IL.

Noss, R. F. 1983. A regional landscape approach to maintain diversity. *Bioscience*, **33**(11), 700–706.

1987a. Protecting natural areas in fragmented landscapes. *Natural Areas Journal*: **7**, 2–13.

1987b. Corridors in real landscapes: a reply to Simberloff and Cox. *Conservation Biology*, **1**, 159–64.

1991. Landscape connectivity: different functions at different scales. In W. E. Hudson, editor. *Landscape linkages and biodiversity*, pp. 27–39. Island Press and Defenders of Wildlife, Washington, DC.

1992. The Wildlands Project: land conservation strategy. *Wild Earth* (Special Issue), pp. 10–25.

1993. Wildlife corridors. In D. S. Smith and P. C. Hellmund (eds.), *Ecology of Greenways: Design and Function of Linear Conservation Areas*, pp. 43–68. University of Minnesota Press, St. Paul, MN.

1995. *Maintaining ecological integrity in representative reserve networks*. WWF Canada/WWF United States, Toronto/Washington, DC.

1996. Protected areas: how much is enough? In R. G. Wright (ed.), *National Parks and Protected Areas: Their Role in Environmental Protection*, pp 91–120. Blackwell Science, Cambridge, MA.

Noss, R. F. and Cooperrider, A. Y. 1994. *Saving Nature's Legacy: Protecting and Restoring Biodiversity*. Defenders of Wildlife, Island Press, Washington, DC.

Noss, R. F. and Harris, L. D. 1986. Nodes, networks, and MUMs: preserving diversity at all scales. *Environmental Management*, **10**(3), 299–309.

Noss, R. F., LaRoe, E. T. III and Scott, J. M. 1995. *Endangered ecosystems of the United States: a preliminary assessment of loss and degradation*. Biological report 28.

National Biological Service, Washington, DC.

Noss, R. F., O'Connell, M. A. and Murphy, D. D. 1997. *The science of conservation planning*. Island Press, Washington, DC.

Nowicki, P. 1998. *Selection of Ecosystems, Types of Habitats and Landscapes of European Importance (Project STRA-REP 11). Project Concept*. Council of Europe, Committee of Experts for the European Ecological Network. STRA-REP (98) 12.

Nowicki, P., Bennett, G., Middleton, D., Reintjes, S. and Walters, R. (eds.) 1996. *Perspectives on Ecological Networks*. ECNC, Tilburg.

Ntiamoa-Baidu, Y., Zéba, S., Mboje Gamassa, D. and Bonnéhin, L. 2000. *Principles in practice: Staff observations of conservation projects in Africa*. USAID, Biodiversity Support Program, Washington, DC.

OECD 1986. *Economic Design of Low-Traffic Roads*. OECD Paris.

Oldfield, S. 1988. *Buffer zone management in tropical moist forest: Case studies and guidelines*. IUCN, Gland.

Opdam, P. 1990. Dispersal in fragmented populations: the key to survival. In R. G. H. Bunce and D. C. Howard (eds.), *Species Dispersal in Agricultural Habitats*, pp. 3–17. Belhaven Press, London and New York.

1991. Metapopulation theory and habitat fragmentation: a review of holarctic breeding bird studies. *Landscape Ecology*, **5**, 93–106.

Opdam, P., Verboom, J. and Pouwels, R. 2003. Landscape cohesion: an index for the conservation potential of landscapes for biodiversity. *Landscape Ecology*, **18**(2), 113–26.

Opstal, A. van 1999. *The architecture of the Pan European ecological network*. National Reference Centre for Nature Conservation. Report No. 37. Wageningen.

Orfield, M. 1997. *Metropolitics: A Regional Agenda for Community and Sustainability*. Brookings Institute, Washington, DC., and Lincoln Institute of Land Policy, Cambridge, MA.

Ostrum, V., Tiebout, C. M. and Warren, R. 1961. The organisation of government in metropolitan areas: A theoretical inquiry. *The American Political Science Review*, **55**, 831–42.

Påhlsson, L. (ed.) 1998. *Vegetationstyper i Norden*. TemaNord 1998: 510. Nordisk Ministerråd, Copenhagen.

Peck, S. 1998. *Planning for Biodiversity*. Island Press, Washington, DC.

Pettingill, O. S. Jr 1970. *Ornithology in Laboratory and Field*. Burgess Publishing Company, Minneapolis, MN.

Pinay, G., Décamps, H., Chauvet, E. and Fustec, E. 1990. Functions of Ecotones in Fluvial Systems. In R. J. Naiman and H. Décamps (eds.), *The Ecology and Management of Aquatic-Terrestrial Ecotones*, Man and the Biosphere Series, Volume 4, pp. 141–69. UNESCO, Paris; and Parthenon Publishing Group, Carnforth.

Platt, W. J. 1998. Southeastern pine savannas. In R. C. Anderson, J. S. Fralish and J. M. Baskin (eds.), *Savannas, Barrens, and Rock Outcrop Plant Communities of North America*, pp. 23–51. Cambridge University Press, Cambridge, UK.

Porter, D. A. 1995. A Fifty Year Plan for Metropolitan Portland. *Urban Land*, **7**, 37–40.

Prat, N., Toja, J., Solá, C., Burgos, M. D., Plans, M. and Rieradevall, M. 1999. Effect of dumping and cleaning activities on the aquatic ecosystems of the Guadiamar river following a toxic flood. *The Science of the Total Environment*, **242**, 231–48.

Prat, N., Solá, C., Plans, M., Toja, J. and Burgos, M. D. 2001. La restauración del estado ecológico del río Guadiamar. *Medio Ambiente*, **36**, 50–5.

President's Commission on Americans Outdoors. 1987. Report and recommendations. Reprinted as: *Americans Outdoors: The Legacy, the Challenge*. US Government Printing Office, Washington, DC.

Press, D. 1998. Local environmental policy capacity: A framework for research. *Natural Resources Journal*, **38** (winter), pp. 29–52.

1999. Local open-space planning in California. In D. A. Mazmanian and M. E. Kraft, (eds.), *Toward Sustainable Communities: Transition and Transformations in Environmental Policy*, pp. 153–83. MIT Press, Cambridge, MA.

Pressey, R. L., Possingham, H. P. and Day, J. R. 1997. Effectivenes of alternative heuristic algorithms for identifying indicative minimum requirements for conservation reserves. *Biological Conservation*, **80**, 207–19.

Provincie Gelderland 1996. *Streekplan Gelderland 1996. Stimulans voor ontwikkeling, ruimte voor kwaliteit en zorg voor omgeving.* Provincie Gelderland, Arnhem.

1997. *Groene connecties. Achtergronden over ecologie, economie en bestuur.* Provincie Gelderland, Arnhem.

1999. *Gelderse Groene connecties. Een gezamenlijke visie van provincie, gemeenten en waterschappen in Gelderland op de uitvoering van ecologische verbindingszones concept.* Provincie Gelderland, Arnhem.

Pungetti, G. 1991. *Water Environment Landscape: A comparison between Dutch and Italian planning.* Pitagora, Bologna.

1995. Anthropological approach to agricultural landscape history in Sardinia. *Landscape and Urban Planning*, **31**, 47–56.

1996. *Landscape in Sardinia: History Features Policies.* CUEC, Cagliari.

1999. State of the art information on ecological networks in Europe and Italy. In A. Farina (ed), *Perspectives in Ecology: A glance from the VII International Congress of Ecology (INTECOL)*, pp. 445–52. Backhuys Publishers, Leiden.

2001. The link between local communities and scientific experts in the framework of regional planning questions in Italy. In *2nd International Symposium of the Pan-European Ecological Network: The partnership of local and regional authorities in the conservation of biological and landscape diversity*, pp. 119–25. Council of Europe, Strasbourg.

Quayle, M. 1995. Urban greenways and public ways: Realizing public ideas in a fragmented world. *Landscape and Urban Planning*, **33**, 461–75.

1996. Urban Greenways and public ways: realising public ideas in a fragmented world. In J. G. Fabos and J. Ahern (eds.), *Greenways: the Beginning of an International Movement*, pp. 461–75. Elsevier, Amsterdam.

Querol, X., Alastuey, A., García Sánchez, A. and López, F. A. (1998). *Seguimiento de la meteorización y toxicidad de los lodos. Jornadas científicas para analizar los resultados obtenidos durante el seguimiento del efecto del vertido tóxico en el entorno de Doñana.* El Rocío, Almonte (Huelva).

Rabenswaaij, C. W. van, Reyrink, L. A. F., Smeets, P. J. A. M. and Sprangers, J. T. C. M. 1991. *Ecologische effecten van aangepaste landbouw.* Rapport Centrum voor Agrobiologisch Onderzoek, Wageningen.

Raik, A. (ed.) 1970. *Scheme of the Complex Territorial Planning of Estonia. Natural Conditions.* Tallinn.

Ralls, K. and Taylor, B. L. 1997. How viable is Population Viability Analysis? In S. T. A. Pickett, R. S. Ostfeld, M. Shachak and G. E. Likens (eds.), *The Ecological Basis of Conservation: Heterogeneity, Ecosystems, and Biodiversity*, pp. 228–35. Chapman and Hall, New York.

Reboratti, C. (ed.). 1989. *La frontera agropecuaria en el Umbral al Chaco.* Desarrollo y perspectivas. Instituto de Geografma, Facultad de Filosofma y Letras. Universidad de Buenos Aires.

Reed, N. 2000. *Comments on Florida growth management review process.* (www.1000friendsofflorida.org)

Reijnen. M. J. S. M., Veenbaas, G. and Foppen, R. P. B. 1995. *Predicting the effects of motorway traffic on breeding bird populations.* Road and Hydraulic Engineering Division/ IBN-DLO, Delft/ Wageningen.

Reijnen, R. and Koolstra, B. 1998. *Evaluatie van de ecologische verbindingszones in de provincie Gelderland.* IBN-DLO, Wageningen.

Remm, K. 2000. Database of Estonian square kilometres. In T. Frey (ed.), *Problems of Contemporary Ecology.* pp. 241–7. Tartu.

Ribeiro, L. F. 1998. *The Cultural Landscape and the Uniqueness of Place: A Greenway Heritage Network for Landscape Conservation of Lisbon Metropolitan Area.* Doctoral Dissertation, Department of Landscape Architecture and Regional Planning, University of Massachusetts, Amherst, MA.

Ricklefs, R. E. 1990. *Ecology*, third edition. W. H. Freeman and Company, New York.

Rientjes, S. and Drucker, G. 1996. An introduction to the Pan-European Biological and Landscape Diversity Strategy – background, philosophy and summary. In: *The Pan-European Biological and Landscape Diversity Strategy*, pp. 10–14. Council of Europe, UNEP, ECNC.

Robertson, G., Mash, M., Tickner, L., Bird, J., Curtis, B. and Putnam, T. (eds.) 1996. *Future Natural: Nature, Science, Culture*. Routledge, London.

Rodoman, B. B. 1974. Polarization of landscape as a manage agent in protection of biosphere and recreational resources. In *Resursy, Sreda, Rasselenije*, pp. 150–62. Nauka, Moscow.

Rogers, C. M. and Caro, M. J. 1998. Song sparrows, top carnivores and nest predation: A test of the mesopredator release hypothesis. *Oecologia, Berlin*, **116**(1–2), 227–33.

Romano, B. 1999. La continuità ambientale nella pianificazione. *Urbanistica*, **112**, 156–60.

2000. *Continuità Ambientale*. Andromeda, Teramo.

Roof, J. and Wooding, J. 1996. *Evaluation of SR 46 wildlife crossing*. Florida Cooperative Fish and Wildlife Research Unit, USBS. Technical Report No. 54. FLDOT FL-ER-61-96.

Roux, A. L., Bravard, J.-P., Amoros, C. and Patou, G. 1989. Ecological changes in the upper Rhône River since 1750. In G. E. Petts, H. A. L. Möller and A. L. Roux (eds.), *Historical Change of Large Alluvial Rivers: Western Europe*, pp. 323–50. John Wiley & Sons, Chichester.

Roy, P. S. and Tomar, S. 2000. Biodiversity characterization at landscape level using geospatial modelling technique. *Biological Conservation*, **95**, 95–109.

Royal Commission on the Future of the Toronto Waterfront 1992. *Regeneration: Toronto's Waterfront and the Sustainable City*. Royal Commission on the Future of the Toronto Waterfront, Toronto, Ontario.

Ryan, K. 1993. *Trails for the Twenty-First Century: Planning, Design, Management Manual for Multi-Use Trails*. Island Press, Washington, DC.

Sabo, P., Koreň, M. Sr, Šteffek, J., Ružičková, J., Koreň, M. Jr, Kramárik, J., Maglocký, Š. and Straka, P. 1996. The Slovak approach to ecological networks. In P. Nowicki, G. Bennett, D. Middleton, S. Rientjes and R. Wolters (eds.), *Perspectives on Ecological Networks*, pp. 19–29. ECNC, Tilburg.

Sanger, M. B. and Levin, M. A. 1992. Using old stuff in new ways: Innovation as a case of evolutionary tinkering. *Journal of Policy Analysis Management*, **11**(1), 88–115.

Saunders, D. A. and Hobbs, R. J. (eds.), 1991. *Nature Conservation*, vol. 2, *The Role of Corridors*. Surrey Beatty & Sons, Chipping Norton, NSW Australia.

Sax, J. L. 1991. Ecosystems and property rights in greater Yellowstone: the legal system in transition. In R. B. Keiter and M. S. Boyce (eds.), *The Greater Yellowstone Ecosystem*, pp. 77–84. Yale University Press, New Haven, CT.

Schaefer, J. M. and Brown, M. T. 1992. Designing and Protecting River Corridors for Wildlife. *Rivers*, **3**(1), 14–27.

Schaefer, J. M. and Smith, D. J. 2000. *Ecological characterization of identified high priority highway-ecological interface zones including the inventory and evaluation of existing Florida Department of Transportation highway facilities within these zones*. Report for Project No. B-B120, Task 1 (Amendment No. 3). Florida Department of Transportation, Tallahassee, FL.

Schreiber, K. F. (ed.) 1988. *Connectivity in Landscape Ecology. Proceedings of the 2nd International Seminar of the International Association for Landscape Ecology. Munsterische Geographische Arbeiten 29*. Munster.

Schwarz, L. L. (ed.), 1993. *Greenways: A guide to Planning, Design and Development*. Island Press, Washington, DC.

Scott, J. M., Davis, F., Csuti, B., Noss, R. F., Butterfield, B., Groves, C., Anderson, J. *et al.* 1993. Gap analysis: a geographic approach to protection of biological diversity. *Wildlife Monographs*, **123**, 1–41.

Searns, R. M. 1995. The evolution of greenways as an adaptive urban landscape form. *Landscape and Urban Planning*, **33**, 65–80.

1996. The evolution of greenways as an adaptive urban landscape form. In:

J. Fabos and J. Ahern (eds.), *Greenways: the Beginning of an International Movement* pp. 65–80. Elsevier, Amsterdam.

Seelye, K. May 12, 2001. Bush Is Choosing Industry Insiders to Fill Several Environmental Positions. *New York Times.*

Seidman, H. 1975. *Politics, Position and Power*. Oxford University Press, New York.

Shafer, C. L. 1991. *Nature Reserves, Island Theory and Conservation Practice* Smithsonian Institution Press, Washington, D.C. and London.

1995. Values and shortcomings of small reserves. *BioScience*, **45**, 80–8.

1999. National park and reserve planning to protect biological diversity: some basic elements. *Landscape and Urban Planning*, **44**, 123–53.

Shaffer, G. B. 1981. Minimum population size for species conservation. *BioScience*, **31**, 131–3.

Shaffer, M. L. 1987. Minimum Viable Populations: coping with uncertainty. In M. E. Soulé (ed.), *Viable Populations for Conservation*, pp. 69–86. Cambridge University Press, Cambridge, UK.

Siegel, F. 1999. Is Regional Government the Answer? *The Public Interest*, Fall 1999, pp. 85–94.

Simberloff, D. 1998. Flagships, umbrellas, and keystones: is single-species management passé in the landscape area? *Biological Conservation*, **83**, 247–57.

Simberloff, D. and Cox, J. 1987. Consequences and Costs of Conservation Corridors. *Conservation Biology*, **1**, 63–71.

Simberloff, D., Farr, J. A., Cox, J. and Mehlman, D. W. 1992. Movement corridors; conservation bargains or poor investments? *Conservation Biology*, **6**, 493–504.

Singer, F. J. and Doherty, J. L. 1985. Managing mountain goats at a highway crossing. *Wildlife Society Bulletin*, **13**, 469–77.

Sluis, T. van der and Chardon, J. P. 2001. How to define European ecological networks. In Y. Villacampa, C. A. Brebbia, and J.-L. Usó (eds.), *Proceedings Ecosystems and Sustainable Development ECOSUD III*, Alicante, Spain, pp. 119–28. Wessex Institute of Technology, Southampton.

Sluis, T. van der, Pedroli, B. and Kuipers, H. 2001. *Corridors for LIFE. Ecological Network Analysis Regione Emilia-Romagna – the plains of Provincia di Modena & Bologna*. Alterra report 365, Wageningen.

Smith, D. J. 1996. *The direct and indirect impacts of highways on the vertebrates of Payne's Prairie State Preserve*. Technical report, Florida Department of Transportation, Tallahassee, FL.

1999a. *Highway-wildlife relationships (Development of a decision-based wildlife underpass road project prioritization model on GIS with statewide application)*. Technical report, Florida Department of Transportation, Tallahassee, FL.

1999b. Identification and prioritization of ecological interface zones on state highways in Florida. In G. Evink, P. Garrett and D. Zeigler, (eds.), *Proceedings of the Third International Conference on Wildlife Ecology and Transportation*, Sept. 13–16, 1999 in Missoula, MT, pp. 209–29. Florida Department of Transportation, Tallahassee, FL.

Smith, D. S. 1993. Greenway case studies. In D. S. Smith and P. Hellmund (eds.), *Ecology of Greenways: Design and Function of Linear Conservation Areas*, pp. 161–208. University of Minnesota Press, Minneapolis, MN.

Smith, D. S. and Hellmund, P. C. (eds.) 1993. *Ecology of Greenways: Design and Function of Linear Conservation Areas*. University of Minnesota Press, Minneapolis, MN.

Snyder, J. R., Herndon, A. and Robertson, W. B. Jr. 1990. South Florida rockland. In R. L. Myers and J. J. Ewel (eds.), *Ecosystems of Florida*, pp. 230–76. The University of Central Florida Press, Orlando, FL.

Soulé, M. E. (ed.) 1987. *Viable Populations for Conservation*. Cambridge University Press, Cambridge, UK.

1991a. Theory and strategy. In W. E. Hudson, (ed.), *Landscape Linkages and Biodiversity*, pp. 91–104. Defenders of Wildlife, Island Press, Washington, DC.

1991b. Land Use Planning and wildlife maintenance: guidelines for conserving wildlife in an urban landscape. *Journal of the American Planning Association*, **3**, 313–23.

Soulé, M. E. and Terborgh, J. (eds.) 1999a. *Continental conservation. Scientific Foundation of Regional Reserve Networks*. Island Press, Washington, DC.

1999b. Conserving nature at regional and continental scales – a scientific program for North America. *BioScience*, **49**, 809–17.

Southall, P. D. 1991. *The relationship between wildlife and highways in the Payne's Prairie basin*. Florida Department of Transportation, Lake City, FL.

Spackman, S. C. and Hughes, J. W. 1995. Assessment of minimum stream corridor width for biological conservation: species richness and distribution along mid-order streams in Vermont, USA. *Biological Conservation*, **71**(3), 325–32.

Speight, M. C. D. 2000. Some thoughts on corridors and invertebrates: the hoverfly (Diptera: Syrphidae) fauna of Abies/Picea forests in temperate west/central Europe. In *Atelier sur les corridors écologiques pour les Invertébrés: stratégies de dispersion et de recolonisation dans le paysage agrosylvicole moderne*, Neuchâtel, mai 2000. Actes des Rencontres Environnement no. 45.

Spellerberg, I. F. 1998. Ecological effects of roads and traffic: a literature review. *Global Ecology and Biogeography Letters*, **7**(5), 317–33.

Spinetti, M. 1997. *L'Aquila Reale, Biologia, Etologia e Conservazione*. Cogecstre, Pescara.

Stenseth, N. C. and Lidicker, W. Z. Jr. (eds.) 1992. *Animal Dispersal: Small Mammals as a Model*. Chapman and Hall, New York.

Stiles, J. H. and Jones, R. H. 1998. Distribution of the red imported fire ant, *Solenopsis invicta*, in road and powerline habitats. *Landscape Ecology*, **13**, 335–46.

Suarez, A., Pfenning, K. and Robinson, S. 1997. Nesting success of a disturbance-dependent songbird on different kinds of edges. *Conservation Biology*, **11**(4), 928–35.

Taylor, J., Paine, C. and FitzGibbon, J. 1995. From greenbelt to greenways: Four Canadian case studies. *Landscape and Urban Planning*, **33**, 47–64.

Telles, G. R. 1975. *Uns Comem os Fogos*. Seara Nova, Lisboa.

1997. *Plano Verde De Lisboa*. Colibri, Lisboa.

Ter Braak, C. J. F., Hanski, I. and Verboom, J., 1998. The incidence function approach to modelling of metapopulation dynamics.

In J. Bascomte and R. Solé (eds.), *Modeling Spatiotemporal Dynamics in Ecology*, pp. 167–88. Springer Verlag and Landes Bioscience, Austin, TX.

Tewes, M. E. and Blanton, D. R. 1998. Potential impacts of international bridges on ocelots and jaguarundis along the Rio Grande wildlife corridor. In G. Evink, P. Garrett and D. Zeigler (eds.), *Proceedings of the Second International Conference on Wildlife Ecology and Transportation*, Feb. 10–12, 1998 in Ft. Myers, FL, pp. 135–9. Florida Department of Transportation, Tallahassee, FL.

The Concise Oxford Dictionary of Current English 1995. Ninth edition. Oxford University Press, Oxford.

The Conservation Fund 2000. *Common Ground*, Vol. 11, No. 2. January/February 2000.

Tilman, D., May, R. M., Lehman, C. L. and Nowak, M. A. 1994. Habitat destruction and the extinction debt. *Nature*, **371**, 65–6.

Tolley, R. S. and Turton, B. J. 1995. *Transport Systems, Policy and Planning: a Geographical Approach*. Longman, Harlow.

Townsend, C. T. and Riley, R. H. 1999. Assessment of river health: accounting for perturbation pathways in physical and ecological space. *Freshwater Biology*, **41**, 393–405.

Trust for Public Land. 1995. *Chattanooga Greenways*. Trust for Public Land, Washington, DC.

Trzcinski, M. K., Fahrig, L. and Merriam, G. 1999. Independent effects of forest cover and fragmentation on the distribution of forest breeding birds. *Ecological Applications*, **9**(2), 586–93.

Turner, M. G. 1989. Landscape ecology: The effect of pattern on process. *Annual Review of Ecological Systematics*, **20**, 171–97.

Turner, T. 1995. Greenways, blueways, skyways and other ways to a better London. *Landscape and Urban Planning*, **33**, 269–82.

US Census Bureau 2000. *Resident Population and Apportionment of the U.S. House of Representatives*. (http://www.census.gov/main/www/cen2000.html)

2001. *Resident Population statistics for Florida*.

US Fish and Wildlife Service 1995. *Second revision Florida panther recovery plan*. Atlanta, GA.

Vannote, R. L., Minshall, G. W., Cummins, K. W., Sedell, J. R. and Cushing, C. E. 1980. The river continuum concept. *Canadian Journal of Fisheries and Aquatic Sciences*, **37**, 130–37.

Veenbaas, G. and Brandjes, J. 1999. Use of fauna passages along waterways under highways. In G. Evink, P. Garrett and D. Zeigler (eds.), *Proceedings of the Third International Conference on Wildlife Ecology and Transportation*, Sept. 13–16, 1999 in Missoula, MT, pp. 253–8. Florida Department of Transportation, Tallahassee, FL.

Velasco, E., Sánchez, M. and Dolz, J. 1991. In J. V. Giráldez *et al.* (eds.), Análisis de la hidrología superficial de las cuencas vertientes a la marisma del Coto de Doñana. *III Simposio sobre el Agua en Andalucía*, vol. 1, pp. 243–56. ITGE, Córdoba.

Verboom, J. 1994. *Een modelstudie naar de effecten van infrastructuur op dispersiebewegingen van dieren*. Rijkswaterstaat/ IBN-DLO, Delft/ Wageningen.

Verboom, J., Foppen, R. Chardon, P. Opdam, P. and Luttikhuizen, P. 2001. Introducing the key patch appoach for habitat networks with persistent populations: an example for marshland birds. *Biological Conservation*, **100**, 89–101.

Verboom, J., Metz, J. A. J. and Meelis, E., 1993. Metapopulation models for impact assessment of fragmentation. In C. C. Vos and P. Opdam (eds.), *Landscape Ecology of a Stressed Environment*, pp. 172–91. IALE Studies in Landscape Ecology 1. Chapman and Hall, London.

Verboom, J., Schotman, A., Opdam, P. and Metz, J. A. J. 1991. European Nuthatch metapopulations in a fragmented agricultural landscape. *Oikos*, **61**, 149–156.

Verkaar, H. J. 1988. The possible role of road verges and river dykes as corridors for the exchange of plant species between natural habitats. In R. F. Schreiber (ed.), *Connectivity in Landscape Ecology*, *Proceedings of the 2nd International Seminar of the International Association for Landscape Ecology*, pp. 79–84. *Munsterische Geographische Arbeiten* 29. Munster.

Vermeulen, H. J. W. 1994. Corridor function of a road verge for dispersal of stenotopic heathland ground beetles (Carabidae). *Biological Conservation*, **3**, 339–49.

1995. *Road side verges: habitat and corridor for carabid beetles of poor sandy soils and open areas*. Wageningen Agricultural University, Wageningen.

Veronica, 1996. *An Ecological Network in the Czech Republic*. (Special Issue.) Brno.

Villano, D. 1993. New hope for the panther. *Defenders*, Spring 1993, pp. 29–35.

Vos, C. C., Baveco, H. and Grashof-Bokdam, C. J. 2002. Corridors and species survival. In K. J. Gutzwiller (ed.), *Applying Landscape Ecology in Biological Conservation*, pp. 84–104. Springer–Verlag, New York.

Vos, C. C. and Chardon, W. J. 1994. *Herpetofauna en verkeerswegen: een literatuurstudie*. Rijkswaterstaat/ IBN-DLO, Delft/ Wageningen.

Vos, C. C. and Opdam, P. (eds.) 1993. *Landscape Ecology of a Stressed Environment*. Chapman and Hall. London.

Vos, C. C., Verboom, J., Opdam, P. and Ter Braak, C. J. F. 2001. Towards ecologically scaled landscape indices. *American Naturalist*, **183**, 24–41.

Vrijlandt, P. and Kerkstra, K. 1994. A strategy for ecological and urban development. In T. Cook and H. N. van Lier (eds.), *Landscape Planning and Ecological Networks*, pp. 71–88. Elsevier, Amsterdam.

Walker, B. (ed.) 1997. *The Reader's Digest Illustrated Great World Atlas*. Reader's Digest Association, Inc. Pleasantville, NY.

Wallin, D. 1995. *Resident use and perception of the Chicago and Calumet Rivers*. Chicago Rivers Demonstration Project, Chicago, IL.

Waterfront Regeneration Trust 1995. *Lake Ontario Greenway Strategy*. Waterfront Regeneration Trust, Toronto, Ontario.

Weiner, D. R. 1988. *Models of Nature*. Indiana University Press, Bloomington, IN.

Wells, M. and Brandon, K. 1992. *People and Parks. Linking Protected Area Management with Local Communities*. World Bank, The World Wildlife Fund, and US Agency for International Development, Washington, DC.

Wiebe, K. 1997. Managing public and private land through partial interests. *Contemporary Economic Policy*, **15**(2), 35–43.

Wilcox, B. A. and Murphy, D. D. 1985. Conservation Strategy: The Effects of Fragmentation on Extinction. *American Naturalist*, **125**, 879–87.

Wilcox, D. A. 1989. Migration and control of purple loosestrife (*Lythrum salicaria* L.) along highway corridors. *Environmental Management*, **13**(3), 365–70.

Wilson, E. O. 2000. A personal brief for The Wildlands Project. *Wild Earth*, **10**(1), 1–2.

Word, R. July 4, 1999. 'Quiet time' is over for St. Joe. *Gainesville Sun*, Gainesville, FL.

Work Programme of the Pan–European Biological and Landscape Diversity Strategy 2001. *Council for the Pan–European Biological and Landscape Diversity Strategy, Fifth Meeting, Strasbourg, 10–11 May 2001.* STRA. CO (2001)1.

Worton, B. J. 1989. Kernel methods for estimating the utilization distribution in home-range studies. *Ecology*, **70**(1), 164–8.

Wright, D. F. 1977. A site evaluation scheme for the use in the assessment of potential nature reserves. *Biological Conservation*, **11**, 293–305.

Wright, G. M., Dixon, J. D. and Thompson, B. H. 1933. *Fauna of the National Parks of the United States.* Volume 1. Government Printing Office, Washington, DC.

Wright, G. M. and Thompson, B. H. 1934. *Fauna of the National Parks of the United States.* Volume 2. Government Printing Office, Washington, DC.

1935. *Fauna of the National Parks of the USA: Wildlife Management in the National Parks.* Fauna Series 2, US Government Printing Office.

Wright, J. B. 1994. Designing and applying conservation easement (Planner's Notebook). *Journal of the American Planning Association*, **60**(3), 380–8.

Yanes, M., Velasco, J. M. and Suarez, F. 1995. Permeability of roads and railways to vertebrates: the importance of culverts. *Biological Conservation*, **71**(3), 217–22.

Zadelhoff, E. van and Lammers, W. 1995. The Dutch ecological network. In G. H. P. Arts, M. van Buuren, R. H. G. Jongman, P. Nowicki, D. Wascher and I. H. S. Hoek (eds.) *Ecological Networks. Landschap* (Special Issue), **95**(3), pp. 77–88. European Centre for Nature Conservation, Tilburg.

Zaitzevsky, C. 1982. *Frederick Law Olmsted and the Boston Park System.* Harvard University Press, Cambridge, MA.

Zube, E. H. 1995. National Parks and Protected Landscapes Planning: From Islands to Regions. In M. Yokohari, S. Yujoba, and K. Shinozawa (eds.), *Ecological Landscape Planning*, pp. 110–16. Process Architecture, Tokyo.

1996. Greenways and the U.S. National Park System In J. Fabos and J. Ahern (eds.), *Greenways: the Beginning of an International Movement*, pp. 17–25. Elsevier, Amsterdam.

Index

Page numbers in italics refer to figures, bold to tables, and in ordinary type to text and boxes.